AF572328

Air Pollution:
Development at What Cost ?

Air Pollution:
Development at What Cost ?

Editors
Yogesh T. Jasrai
&
Arun Arya
Department of Botany, Faculty of Science,
The Maharaja Sayajirao University of Baroda,
Vadodara - 390002

2003

DAYA PUBLISHING HOUSE
Delhi - 110 035

ISBN 81-7035-282-7

Published by : **Daya Publishing House**
1123/74, Deva Ram Park,
Tri Nagar, Delhi - 110 035
Phone : 7103999
Fax: (011) 7199029
e-mail: dayabooks@vsnl.com
website: www.dayabooks.com

Showroom : 4762-63/23, Ansari Road, Darya Ganj,
New Delhi - 110002
Phone: 3245578, 3244987

Laser Typesetting : **Classic Computer Services**
Delhi - 110 035

Printed at : **Chawla Offset Printers**
Delhi - 110 052

PRINTED IN INDIA

Foreword

The twin problems of environmental degradation and pollution, in a developing country like India, have started assuming alarming proportions. Our rural environment is degrading rapidly, due to over-exploitation of natural resources, whereas our urban complexes are suffering from a variety of pollution problems, including rapidly spreading urban slums. The quality of life for the people of the country is on the decline.

It is this context, this edited volume on air pollution issues becomes significant. Any contribution, like this volume which covers a whole range of pollution related topics would help in creating environmental awareness amongst a whole range of factors, concerned with this problem. More importantly, a synthesis of available information should help the scientific community in identifying the gap areas for research. Covering a whole range of topics, this is a welcome addition to the Indian literature on the subject.

P.S. Ramakrishnan
Professor of Ecology
School of Environmental Sciences,
Jawaharlal Nehru University,
New Delhi – 110 067
Ph: (O) 91-11-6107676, 6167557 Ext. 2326, 6172438 (Direct)
Fax: 91-11-6172438, 6165886, 6169962
E-mail: psrama@jnuinv.ernet.in ; psrama2001@hotmail.com

Preface

The problem of air pollution has been a major environmental hazard of ever increasing industrialization and urbanization during the 20th century, affecting adversely both physical as well as biological systems. The unprecedented increase in the number and the activities of human, since the industrial revolution has deteriorated the environment that threatens the future of the planet. Terms like **global warming, greenhouse effect, ozone depletion, bio-diversity** etc. have become familiar even to a common man's ear. The important pollutant gases are O_3, SO_2, NO_X, CO and peroxyacetyl nitrate (PAN), although several other gases such as H_2S, HF, NH_3, CH_4, etc. also acquire importance in some localized areas during episodic and accidental releases.

In the 21st century one question is certain to trouble many minds. What is the price paid in the name of development? No one can put a price tag on the species gone extinct, or natural habitats irreparably destroyed. Yet, it is flair for the money that activates the destruction of the environment. Traditional life style and ancient wisdom, which co-existed with nature, are tragically lost with the present development.

Many basic questions come to our mind. Do we have a choice about the air we breathe, the water we drink, food containing residues of the pesticides we eat, the deafening noise levels we encounter in cities or the trees that are chopped to widen the roads? Do we really need a healthy and harmonious environment around ourselves? Or, we are least worried on such issues.

A **national seminar** was organized by IAAPC on 16th October 1999 in Baroda, to discuss such issues related to **Pollution and Development**. First part of this volume contains original papers presented in the seminar and second part deals with informative invited articles in the field of pollution. The book is unique in the sense that various issues related to air pollution like **sources,**

monitoring, impact assessment and its effect on plant and human health are discussed. The plants absorb large quantities of air pollutants, acting as their natural sinks and thus 'purifying' the air. The suggestion of a '**green belt**' between industries and the residential areas is borne out of this property of the plants. Now it is high time to pay adequate attention for abatement of pollution. These researches will surely help to update our knowledge in the field at the same time efforts made by environmental scientist and technologists pave way for a better environment and sustainable development.

We are thankful to our wives Mrs. S. Jasrai and Dr. (Mrs.) C. Arya for their help, critical suggestions and providing a congenial platform to perform our duties upto the latest satisfaction. One of the authors (AA) is really thankful to Mr. Anshul, his son, who inspired the author to work on the problem.

We are thankful to our President, Prof. V.S. Patel, and to Secretary Prof. S.J. Bedi for their encouragement. We also thank all the contributors responding to submit their manuscript in corrected form. Thanks are also due to Mr. Ravi our photo-artist, and Shri Anil Mittal of Daya Publishing House for nice printing of the volume.

Prof. Y.T. Jasrai
Dr. A. Arya

Contents

List of Contributors

Abraham, Leena
Botany Department, Faculty of Science, The M.S. University of Baroda, Vadodara - 390002

Agrawal, M.
Department of Botany, Banaras Hindu University, Varanasi - 221005

Arya, Arun
Botany Department, Faculty of Science, The M.S. University of Baroda, Vadodara - 390002

Arya, Chitra
Botany Department, Faculty of Science, The M.S. University of Baroda, Vadodara - 390002

Babu, M.K.G.
Center of Energy Studies, I.I.T., Delhi

Bala, S.S.
Central Pollution Control Board, Western Zone, Synergy House, Subhan Pura, Vadodara - 390007

Balagurunathan, K.
Mechanical Engineering Department, College of Engineering, Guindy, Anna University, Chennai - 600025

Bhatt, N.M.
Civil Engineering Department, Faculty of Technology and Engineering, M.S. University of Baroda, Vadodara - 390001

Bhatt, R.D.
Civil Engineering Department, Faculty of Technology and Engineering, M.S. University of Baroda, Vadodara - 390001

Bhavnani, H.V.
Deptt. of Civil Engineering, Faculty of Technology & Engg., M.S. University of Baroda, Vadodara - 390001

Ch. Uma
Department of Biosciences, Sardar Patel University, Vallabh-Vidyanagar - 388120

Chandrasekar, J.
Mechanical Engineering Department, College of Engineering, Guindy, Anna University, Chennai - 600025

Chawda, S.
Civil Engineering Department, Faculty of Technology and Engineering, M.S. University of Baroda, Vadodara - 390001

Dubey, P.S.
Institute of Environment Management and Plant Sciences, Vikram University Ujjain - 456010

G. Prathapsenan

Department of Botany, Faculty of Science, M.S. University of Baroda, Vadodara - 390002

Galgale, A.D.
Civil Engineering Department, Faculty of Technology and Engineering, M.S. University of Baroda, Vadodara - 390001

Garge, Sandhya K.

Department of Botany, Faculty of Science, M.S. University of Baroda, Vadodara - 390002

Inamdar, J.A.
Department of Biosciences, Sardar Patel University, Vallabh-Vidyanagar - 388120

G. Pradeep Kumar
Department of Botany, Faculty of Science, M.S. University of Baroda, Vadodara - 390002

Jain, N.K.
Institute of Environment Management and Plant Sciences, Vikram University, Ujjain - 456010

Jasrai, Y.T.
Department of Botany, Faculty of Science, M.S. University of Baroda, Vadodara - 390002

Kamah, B.V.
Department of Chemistry, Faculty of Science, M.S. University of Baroda, Vadodara - 390002

Kotecha, P.V.
Preventive and Social Medicine Department, Medical College, Vadodara - 390001

Kumar, R.R.
Mechanical Engineering Department, College of Engineering, Anna University, Guindy Campus, Chennai - 600 025

Kumar Sheo
Botanical Survey of India, Kamla Bhawan, P-8, Brabourne Road, Kolkata - 700001

Kumawat, D.M.
Institute of Environment Management and Plant Sciences, Vikram University, Ujjain - 456010

Kumbhkar, K.K.
Institute of Environment Management and Plant Sciences, Vikram University Ujjain - 456010

Mathur, H.B.
Mech. Engg. Dept. D.C.E., Delhi

Mehta, Nitin
Institute of Environment Management & Plant Sciences, Vikram University, Ujjain - 456010

Nansey, Jaidev
3, Patel Colony, Jamnagar - 361008

Nirmal, R.A.
Mechanical Engineering Department, College of Engineering, Guindy, Anna University, Chennai - 600 025

Padate, G.S.
Division of Avian Biology, Department of Zoology, Faculty of Science, M.S. University of Baroda, Vadodara - 390002

Pandya, P.
Institute of Environment Management and Plant Sciences, Vikram University, Ujjain - 456010

Prasad, S.
Mechanical Engineering Department, College of Engineering, Anna University, Guindy Campus, Chennai - 600 025

Patel, V.S.
Visiting Professor, Faculty of Technology and Engineaing, M.S. University of Baroda, Vadodara - 390001

Ramana Rao, T.V.
Department of Biosciences, Sardar Patel University, Vallabh-Vidyanagar - 388120

Ramanathan, R.
Mechanical Engineering Department, College of Engineering, Anna University, Guindy campus, Chennai - 600 025

Rao, P.S.

Executive Director, Gujarat Refinery, P.D. Jawaharnagar - 391320

Rao, S.B.
Mech. Engg. Deptt., Dayalbagh Educational Institute, Dayalbagh, Agra - 282005

Raole,V.M.
Botany Department, Faculty of Science, The M.S. University of Baroda, Vadodara - 390002

Sadasivan, S.
Museum and Picture Gallery, Vadodara - 390018

Sapana, S.
Division of Avian Biology, Department of Zoology, Faculty of Science, M.S. University of Baroda, Vadodara - 390002

Shah, Keyur
Central Pollution Control Board, Western Zone, Synergy House, Subhan Pura, Vadodara - 390007

Shah, A.R.
Museum and Picture gallery Baroda, Vadodara - 390018

Shah, N.R.
Dept. of Museology, Faculty of Fine Arts, M.S. University of Baroda, Vadodara - 390002

Singh, B.
Department of Botany, Faculty of Science, Banaras Hindu University, Varanasi - 221 005

Singh, S.N.
Environmental Science Division, National Botanical Research Institute, Lucknow - 226 001

Singh, Vasdeo
Chemical Engineering Department, Faculty of Technology & Engineering, M.S. University of Baroda, Vadodara - 390002

Thanaraj, K.
Mechanical Engineering Department, College of Engineering, Anna University, Guindy campus, Chennai - 600025

Trivedi, R.C.

Centre for Environment Education, Thaltej Tekra, Ahmedabad - 380054

Vashi, B.D.
Civil Engineering Department , Faculty of Technology & Engineering, M S. University of Baroda, Vadodara - 390001

Section I

Air Pollution: Development at What Cost?

CHAPTER 1

Air Pollution Monitoring

Keyur Shah and S.S. Bala
Central Pollution Control Board
Western Zone, Synergy House,
Subhanpura, Vadodara

Air Pollution

The atmosphere is composed primarily of nitrogen (78.08%), oxygen (20.95%) and many other noble gases, the concentration of which have been found to be fixed over time. When these natural exchange of gases get disturbed by means of man made intervention, the concentration of atmospheric gases may reach to a dangerous level which cause serious health hazards. Unfortunately, the vast modernisation disturbs the atmospheric system of gases which compel us to worry about.

Source of Air Pollution

Sources of air pollution are many and we all are a part of air pollution-causing process in our everyday life. The source can be divided in two broad categories:

1. Natural and
2. Man made

Volcanic eruptions, forest fires, desert storms and biosynthesis may be considered as natural intervention. Natural sources of air pollution are usually beyond the control.

Man-made sources are, generally more prevalent than natural sources and most important with regard to generation of air create air pollution, which can be easily controlled. The man-made intervention can be further divided into the following groups:

1. Industrial
2. Mobile
3. Domestic
4. Non-point

Industries emit pollutant gases to the atmosphere through chimneys. Different types of industries emit different types of pollutants like Cl_2, NH_3, acid mist, heavy metals etc. in varying quantities.

Domestic fossil fuel consumption is the main source of domestic air pollution. However, the intensity of this problem has now reduced due to the switching over of fuel from coal/kerosine/wood to the L.P.G., particularly in urban centres.

The automobiles, which include cars, buses, trucks, tempos etc. are the main mobile pollution sources. The mobile sources are the main air polluters in urban centers particularly with respect to the parameters like NO_X, HC, CO etc.

Air Pollution Control Equipments

There are several devices available for controlling the air pollution emitted from industries. Cyclones, Electrostatic precipitators, Bag filters, Scrubbers, Adsorption device are the equipments normally being used in the industries for controlling air pollution.

Air Pollution Monitoring

The control equipment used to contain the pollution generated from petrol driven vehicle is the catalytic convertor.

Air Quality Monitoring Equipment

There is a need for monitoring air quality to quantify its concentrations and effects upon the health of human beings. Further, the status of air quality is one of the prerequisite data to frame an air quality management programme for a region as a long term measure to control the air pollution in future. The air pollution monitoring is also required to determine the efficacy, efficiency of the system.

Instruments are available to monitor industrial air pollution, mobile air pollution, ambient air quality etc. with a high degree of accuracy and precision.

- Stack monitoring kit used for monitoring the stack gases (source monitoring) from process and utility operations.
- High and low volume sampler, used for monitoring the ambient air quality.
- CO/HC analyzer used for monitoring the exhaust, from petrol driven vehicles.
- Smoke meter used for monitoring the density of smoke emitted from diesel driven vehicle.

Every equipment is developed for monitoring a certain range of concentrations and work under specific conditions. The general specification of stack monitoring kit are:

- Stack velocity range : 0 to 30 m/sec
- Stack temperature range : 0 to 600°C
- Particulate sampling : At 10 to 100 lpm
- Filter paper (Thimble) : Collection of particulate down to 0.3 micron
- Gaseous sampling : At 1 to 2 lpm collection on a set of impingers, containing selective reagents.

Source Monitoring

Monitoring of Particulate Matter

The stack monitoring kit (sink) is commonly used for point source monitoring. The sampling point should be selected in the stack at any cross section that is atleast eight stack diameter downstream and two diameter upstream from any flow disturbances such as bend expansion, contraction or stack exit. The sampling porthol and platform requirements should be as per the guidelines prescribed by Central Pollution Control Board (CPCB) in the document "Emission Regulations Part-III. After selecting the site, the sampling should be carried out at different traverse points under isokinetic condition i.e. kinetic energy of the gas stream in the stack should be equal to the kinetic energy of the gas stream through the

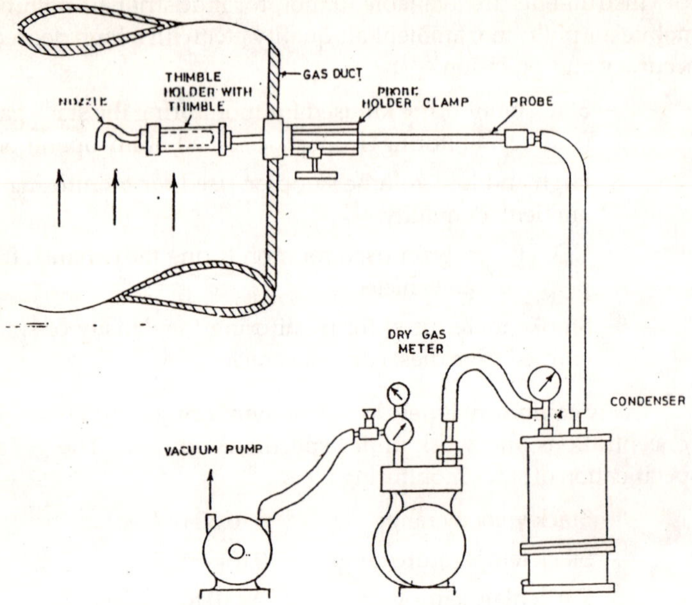

Fig.1: Thimble Sampling Train

sampling nozzle. Figure 1 presents the sampling train for the source monitoring of particulate matter and gaseous substances.

The sampling consists of several distinct steps to obtain the various parameters and finally arrive at the main formula for determining the concentration of stack gaseous and dust concentrations.

The volume of dry gas can be calculated by using the following equation:

$$Vstd = Vm\ Y\ Pbar - Pm/760 \times 273 + 25/Tm + 273$$

Where,

Tm	=	Temperature of gas at dry gas meter condition, °C.
Vm	=	Volume of gas sampled at dry gas meter conditions, m^3.
(Pbar – Pm)	=	Actual pressure in sampling train, mm mercury column

Pbar	=	Barometric pressure, mm mercury column
Pm	=	Static pressure in sampling train, mm mercury column
Y	=	Calibration factor of dry gas meter

Similarly, the dust concentration can be calculated from the equation

Dust concentration in mg/Nm^3 S = Mn/Vstd,
(25°C, 760 mm Hg, dry basis)

Where,

Mn	=	Mass of the dust in the thimble in mg
Vstd	=	Volume of dry gas through the meter (25°C, 760 mm Hg), NM^3

All stack emission test results are reported on dry basis i.e. at zero per cent moisture and for the normalized conditions i.e. 25°C and 760 mm Hg.

Gaseous Source Monitoring

The monitoring of gaseous pollutants like SO_2, NO_X, NH_3, Cl_2 etc. generated from the stack is carried out using either stack monitoring kit or handy samplers or other than detectors.

Ambient Air Quality Monitoring

The high volume sampler is used for measuring suspended particulate matter anlomer gaseous pollutants. A low volume sampler may be used for measuring ambient qualities of gaseous polluters.

The high volume sampler is a monitoring device, complete in all respects including blower, filter hodler, free from leakages, cabinet, roof shelter, automatic voltage stabilizer, flow measurement device and arrangement of gaseous sampling with flow controller.

Blower suck the air with a constant flow rate between 1.4 to 1.8 m^3/min and the particulate matter less than 100 micron are collected on a EPM–2000 Whatman filter paper. The difference of initial and final weights of the filter paper gives the concentration of the particulate matter.

For gaseous sampling, different absorbents are used for absorbing the ambient gases like SO_2, NO_X, Cl_2 etc. Then these absorbent are subject to the analysis.

Selection of Monitoring Locations

This is a very crucial aspect, as the right direction and distance of the ambient air quality monitoring stations would reveal a correct picture of the dispersion of the gaseous pollutants emitted from a sources and thus the environmental impacts of the project activities.

The selection of monitoring location is usually made at 120° around the project site at various radical distance, to cover the entire spectrum of the impact zone. The difference in the air quantities at two locations. One is upstream and another in downstream of project site could easily be correlated with activities at the project site (industry). The height of the monitoring location is usually 10-15 feet. The meteorological parameters which are vital from air pollutants diffusion and dispersion point of view, should be collected at the industry site, usually.

Frequency of Air Quality Monitoring

The frequency of source monitoring and ambient air quality may very depending upon the size and type of industry. Table 1 presents various frequencies to be used to monitoring various types of listed industries.

Table 1: Frequency of Air Quality Monitoring

Plant Capacity TPA	*Ambient Air Quality Monitoring Stations*	*Source Emission Monitoring*
Cement		
Less than 100000 and including mini cement plant	Not required	Once in 8 weeks
100000 upto 300000	2 Stations	Once in 4 weeks
300000 upto 600000	3 Stations	Once in 2 weeks
600000 and above	4 Stations	Once in 1 week
Boiler Capacity TPA	*Ambient Air Quality Monitoring Stations*	*Source Emission Monitoring*
Thermal Power		
Less than 200	2 Stations	Once in 4 weeks
Greater than and Including 200	3 Stations	Once in 2 weeks
Greater than and including 500	4 Stations	Once in 1 week

Contd..........

Integrated Iron and Steel

Plant	*Source Emission Monitoring*
Sintering	Once in 2 weeks
Steel making	Once in 2 weeks

Fertilizer

Plant Type	*Ambient Air Quality Monitoring Stations*	*Source Emission Monitoring*
Phosphatic	Three	Once in 4 weeks
Nitrogenous	Three	Once in 8 weeks
Mixed	Three	Once in 8 Weeks

Nitric Acid

Plant Capacity TPD	*Ambient Air Quality Monitoring Stations*	*Source Emission Monitoring*
Less than 150	2 Stations	Once in 4 weeks
Greater than and including 150	3 Stations	Once in 2 weeks

Sulphuric Acid

Plant Capacity TPD	*Ambient Air Quality Monitoring Stations*	*Source Emission Monitoring*
Less than 100	2 Stations	Once in 4 weeks
Greater than and including 100	3 Stations	Once in 2 weeks

Oil Refinery

Refining Capacity tonnes crude/day	*Ambient Air Quality Monitoring Stations*	*Source Emission Monitoring*
Less than 3000	3 Stations	Once in 8 weeks
3000 and higher	4 Stations	Once in 2 weeks

2

Air Quality in Three Different Cement Producing Areas

Nitin Mehta and P.S. Dubey
Institute of Environment Management & Plant Sciences, Vikram University, Ujjain

Abstract

Ambient air quality in three important cement producing areas was assessed on the basis of SPM, SO_2 and NO_X concentration. Results reveal that SPM is maximum around old established plant like J.K. White Cement Works, while in Vikram Cement which is established after the J.K. White is having a lesser SPM concentration. Of all these the newly established cement plants like Aditya Cement is having a minimum SPM concentration around all the three cement producing areas. These observations directly corresponds to the equipments installed and their performance for dust control in these respective units. In case of SO_2 concentration data reveal the same trend as were recorded for the SPM concentration. But as for as NO_X is concerned it was recorded as maximum at newly established cement plant probably because of the fugitive emission from the highways

Introduction

Air pollution is a major threat at dawn of 21st century. One of the early steps involving air pollution problems is to locate the source from which air contaminants are being emitted. There are more than 7,50,000 man made chemicals present in our environment and to these 1000-2000 new ones are added every year. Massive production of such chemicals directly or indirectly releases thousands of tonnes of a variety of air pollutants into the atmosphere. Some of the air pollutants emanated into the atmosphere are SPM, SO_2 and NO_x etc.

Indian cement industry has been the nation's construction industry for nearly eight decades. From small beginning in the year 1914, this has grown into a major industry in India. By 1999 the total production is likely to exceed 90 million tonnes. With limestone and coal deposits in abundance (limestone- 91 billion tonnes & Coal- 130 billion tonnes) it is poised to grow further to be a major cement producing nation. In cement industries dust is released into the atmosphere, with both visible and latent consequences on ecosystem components and have been a common air pollutant in the vicinity of the manufacturing plant.

The objective of this study is to assess the air quality around different cement producing areas so that comparative data of air quality will be available in three different cement producing areas.

Materials and Methods

After an extensive survey of the area, different sites were selected for monitoring ambient air quality keeping in view the meteorological conditions (Fig. 1, 2, 3). On the basis of survey sampling programme was designed for a period of one year including all seasons. Average values of November, December, January and February were pooled to obtain the mean value of Winter, March, April, May and June to obtain mean value of Summer and July, August, September and October were pooled to get mean value of Rainy Season.

Nayagaon - Khor area lies at the end of Malwa plateau in South-Western M.P. near Rajasthan border. The area has two cement industries, Vikram Cement (VC) and Cement Corporation of India (CCI). It was observed that South-West (SW) wind blows towards North-East (NE) during summer and rainy seasons and during winter season South-East (SE) winds blows towards North-West(NW). In the present study area Shambhupura is located near Chittorgarh district of Rajasthan. The area has Aditya Cement here. Gotan is located in Nagaur district of Rajasthan. The area has four Cement industries i.e. J.K. White Cement Works, Birla White Cement, Snowcem Cement and Neon Cement. In these two areas the wind direction are similar to as were observed in Nayagaon-Khor area. Since it is well established that maximum load of particulate occur in vicinity of the source i.e. 0.5 to 1.5 Km (Agrawal and Agrav. al, 1989), monitoring was mostly conducted in this region.

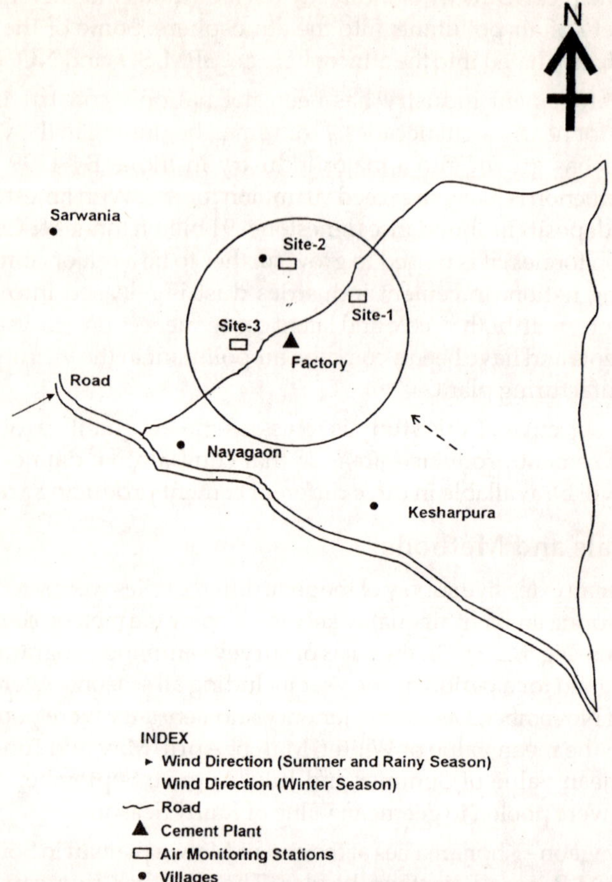

Fig.1 : Location of Sampling site at Nayagaon-Khor area around Vikram Cement

Estimation of Air Quality

Suspended Particulate Matter (SPM) is measured with HVS (120, Kimoto, Japan) using filter paper EPM 2000 at 1.1-1.6 LPM Flow rate and gaseous levels employing Portable Gas Sampler (NETEL'S NPM- PSI) at 0.5 LPM flow rate by using absorbing regent (TCM for SO_2 & NaOH for NO_X). After monitoring the air further analysis is done in the laboratory.

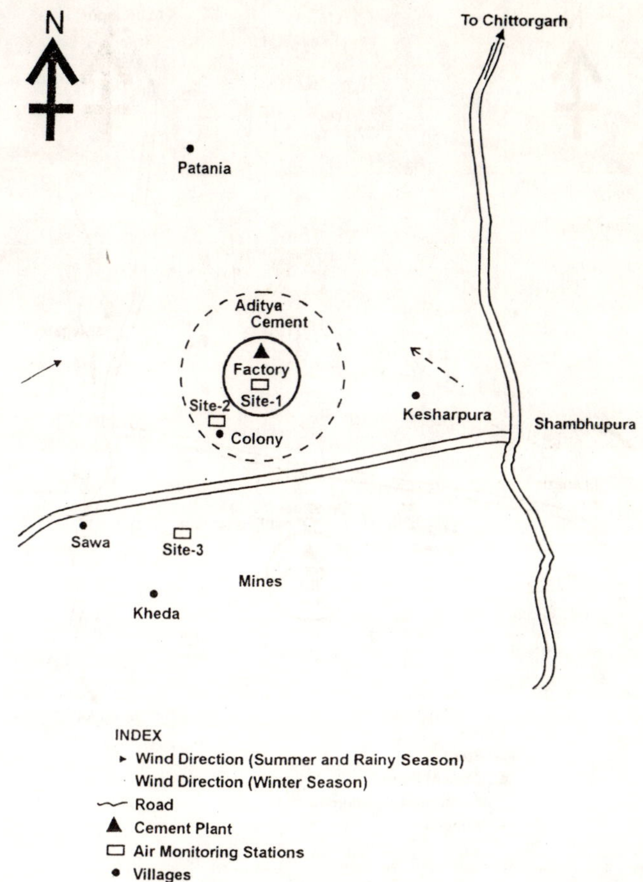

Fig.2 : Location of Sampling site at Sawa-Shambhupura area around Aditya Cement

Results and Discussion

Apparently, the data presented show that these areas are polluted where SPM, SO_2 and NO_X exists in good amount even at 1 Km. distance. This was surprising since in Khor area the cement production is to the tune of 4.5 MT while at Gotan it is 0.18 MT and at Sawa Shambhupura it is 1 MT but ambient air quality is almost

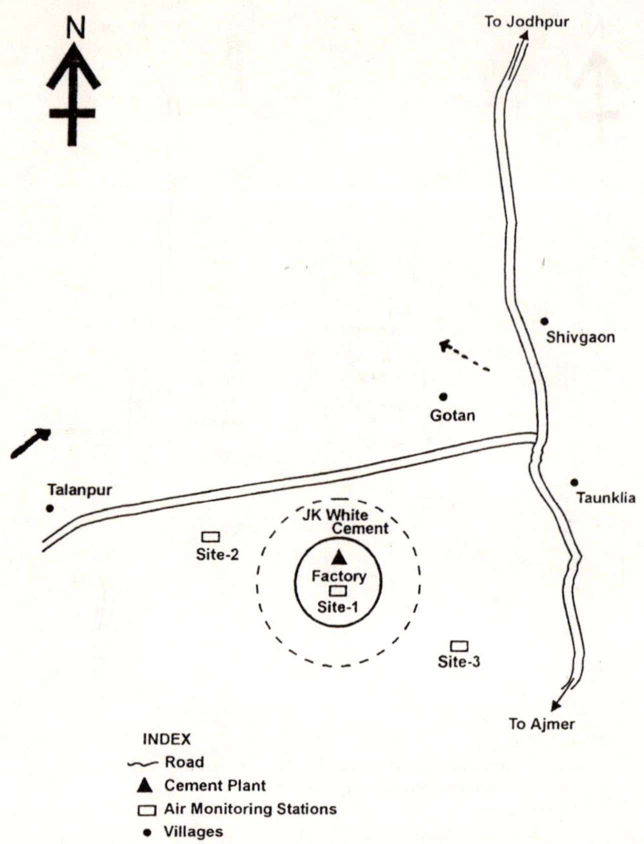

Fig.3 : Location of Sampling site at Gotan area around J.K. White Cement

equivalent. This situation was to be analysed and understood and hence a detailed study was planned under the projects sponsored by these industries. Because the earlier studies of Reddy (1997), Sharma (1996), Mankotia (1996) analysed the environmental quality with simple view points of studying the impact of the cement industries on soil and plants but no attempt was made to really understand the cause of various factors inducing more pollution in the area inspite of the efficient dust pollution control equipments.

Table 1 : SPM Concentration (μg/m³) in the Nayagaon-Khor area around Vikram Cement

Sl.no.	*Season*	*(Distance from the source)*		
		100 meter	*500 meter*	*1000 meter*
1	Summer	244.21	280.70	199.82
2	Rainy	180.08	148.35	151.08
3	Winter	369.59	328.66	176.02
4	Annual Avg.	2664.62	252.57	175.64

Table 2 : SO_2 Concentration (μg/m³) in the Nayagaon-Khor area around Vikram Cement

Sl.no.	*Season*	*(Distance from the source)*		
		100 meter	*500 meter*	*1000 meter*
1	Summer	10.14	13.79	0.98
2	Rainy	13.50	19.79	0.89
3	Winter	20.79	26.57	13.43
4	Annual Avg.	14.81	19.95	5.1

Table 3 : NO_x Concentration (μg/m³) in the Nayagaon-Khor area around Vikram Cement

Sl.no.	*Season*	*(Distance from the source)*		
		100 meter	*500 meter*	*1000 meter*
1	Summer	13.34	7.84	0.67
2	Rainy	5.76	8.48	0.43
3	Winter	13.59	7.19	11.21
4	Annual Avg.	10.89	7.83	4.10

Table 4 : SPM Concentration (μg/m³) in the Sawa-Shambhupura area around Aditya Cement

Sl.no.	*Season*	*(Distance from the source)*		
		100 meter	*500 meter*	*1000 meter*
1	Summer	280.25	1466.70	263.63
2	Rainy	181.77	138.19	248.65
3	Winter	317.08	192.45	288.12
4	Annual Avg.	259.70	159.04	266.80

Table 5 : SO_2 Concentration (μg/m³) in the Sawa-Shambhupura area around Aditya Cement

Sl.no.	Season	(Distance from the source)		
		100 meter	500 meter	1000 meter
1	Summer	14.05	6.70	17.03
2	Rainy	7.93	8.98	12.40
3	Winter	24.0	4.5	26.06
4	Annual Avg.	15.32	4.06	18.48

Table 6 : NO_x Concentration (μg/m³) in the Sawa-Shambhupura area around Aditya Cement

Sl.no.	Season	(Distance from the source)		
		100 meter	500 meter	1000 meter
1	Summer	20.05	9.10	16.12
2	Rainy	24.62	0.88	10.25
3	Winter	24.0	4.5	26.06
4	Annual Avg.	21.59	13.90	6.80

Table 7 : SPM Concentration (μg/m³) in the Gotan area around J.K. White Cement

Sl.no.	Season	(Distance from the source)		
		100 meter	500 meter	1000 meter
1	Summer	310.30	266.39	218.71
2	Rainy	198.21	185.41	147.89
3	Winter	347.28	340.40	175.8
4	Annual Avg.	285.26	264.06	180.80

Table 8 : SO_2 Concentration (μg/m³) in the Gotan area around J.K. White Cement

Sl.no.	Season	(Distance from the source)		
		100 meter	500 meter	1000 meter
1	Summer	53.45	9.20	42.83
2	Rainy	26.44	0.20	19.50
3	Winter	37.83	13.91	50.65
4	Annual Avg.	39.24	7.77	37.66

Table 9 : NO_x Concentration (μg/m³) in the Gotan area around J.K. White Cement

Sl.no.	*Season*	*(Distance from the source)*		
		100 meter	*500 meter*	*1000 meter*
1	Summer	0.029	0.026	0.043
2	Rainy	0.019	0.011	0.020
3	Winter	0.027	0.014	0.036
4	Annual Avg.	0.025	0.017	0.033

The results presented in this paper pertains to the ambient air quality of three areas where cement production is in progress along with production of stones and lime. The three areas are, J.K. White Cement, Gotan (Raj.) where about 1.8 lakh for white cement is produced, other area is Aditya Cement Shambhupura (Raj.) with 1 million tonnes of grey cement and large number of quarries to cut slab stones, exists and third is Nayagaon Khor area, where Vikram Cement with 4.5 million tonnes of cement production. In this area itself is CCI, with a production capacity of 1.5 million tonnes. No lime kilns and stone cutting operations of are underway here. Aditya Cement is a new plant, i.e. 3 years old only while other two plants are about 14-15 years old.

The ambient air quality around Vikram Cement area shows a distinct gradient oriented concentration of SPM along the distance from the point source with an annual average of above 175.64 μg/m³. The levels of SO_2 is quite low with annual average of 5.1 μg/m³ while the NO_x is still lower 4.1 μg/m³.

Around Aditya Cement the situation for SPM levels is not along any gradient in wind direction and interestingly we observed that SPM levels is maximum than the plant premises. The industry has very efficient dust emission control and the higher levels of SPM at 1000 M is 266.80 μg/m³ because of the large numbers of stone quarries and cutting operations which even at 3-4 Km distances has a background SPM load of 150-170 μg/m³ in winters. The levels of SO_2 is in the range of 18.48 μg/m³ (Annual average) at 1000 M distance while the levels of NO_x is in the range of 6.8 (Annual average) at 1000 M distance.

The area around J.K White Cement exhibit higher levels of SPM on seasonal as well as annual basis, increased levels of SO_2 too

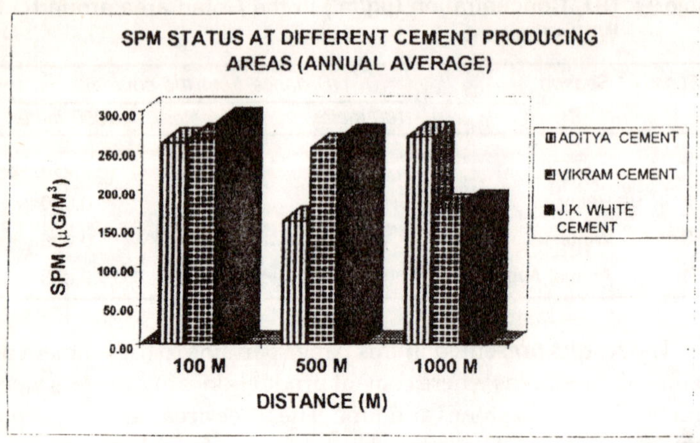

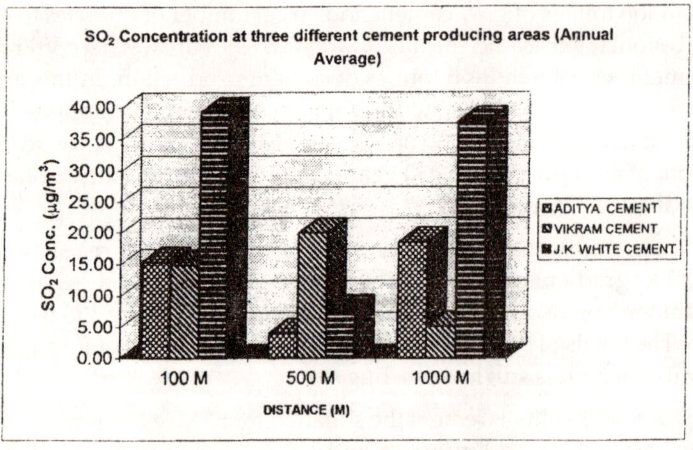

exhibit same trend while level of NO_X is quite low. It is important to note that the J.K. White Cement is a small unit producing hardly 0.18 MT of white cement and has a very efficient pollution control. In fact the area remains charged with these pollutants much because of the famous lime of Gotan produced in a large number of lime kiln around the area.

The main objective of this paper is that the areas around cement production units remain polluted, but the contribution of these units

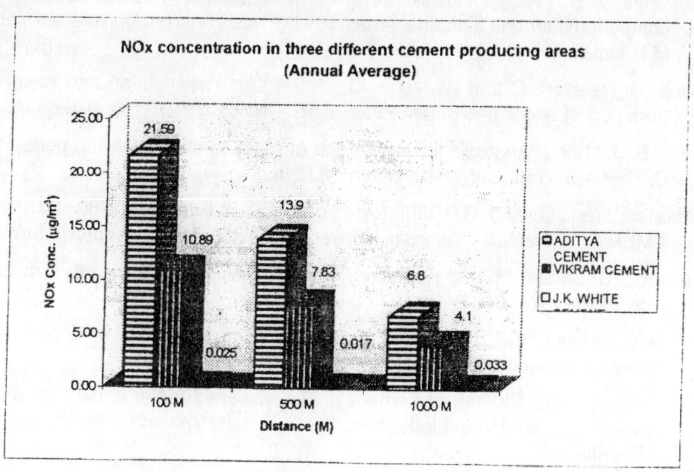

is not very high. For example around Aditya Cement the stone cutting operations, quarrying etc. resulted in a background concentration of nearly 150-170 µg/m^3 even earlier to the installation of the Aditya Cement. In Gotan area it is all the contribution of the lime kiln, while around the Vikram Cement it is because of heavy traffic and a production of about 6.0 MT of cement in a small area.

The tables further reveals that seasonal variations that occur at all sites with a maximum pollution load during winter for all the pollutants because of poor inversion. The ambient air quality as a general rule appears to be better during the rainy seasons because of the rain washings.

References

Agrawal, M. & Khanam, N. (1997). Variations in concentrations of particulate matter around a cement factory. *Indian J. Environ. Health*, 39 (2); 97-102p.

Lerman, S. & Darley, E.F. (1975). Particulates In: Response of Plants to air pollution Eds. : J.B., Mudd and T.T., Kozlowski. Academic Press, NY. : 141-157p.

Soni D.K. and Aggrawal A.L. (1997). Characterisation of dust emission in coal mining activities - Case study. *Ind. J. Environ. Protec.* 17(11). 809-814p.

Leck. C. and Rodhe, H. (1989). 'On the relations between anthropogenic SO_2 emission and concentration of Sulphate in air and precipitation'. *Atmos. Environ.* 23: 959-966p.

McLaughlin, S.B. (1985). 'Effect of air pollution on forests'. *J. Air pollut. Control. Ass.* 35: 512-534p. OECD Environmental Data, Report No. ISBN 92-64-134954-8, pp. 7- 40 Paris (1991).

Mankotiya, A.S. (1996). Heavy metal contaminations in some ecosystem components around industrial areas. Ph. D. Thesis, Vikram University, Ujjain, M.P. India.

Rodhe, H. Persson. C and Akesson. O. (1992). 'An investigation into regional transport of shoot and sulphate aerosols'. *Atmos. Environ.* 6: 675-693p.

Reddy, B.M. (1997). An ecological approach of study to develop green belts. Ph. D. Thesis, Vikram University, Ujjain, M.P., India.

Saltbones, J. and Doland, H (1986). Emission of SO_2 in Europe in 1980 and 1983. EMEP/CCC-Report, Lillestrom, Norwegian Institute for air research, Norway.

Spenjler, J.D. Brauer, M. and Koutrakis, P. (1990). Acid air and health. *Environ. Sci. Technol.* 24: 946-956p.

Stern, A.C., Boubel, R.W., Turner, D.B. and Fox, D.L. in; Fundamental of Air pollution, 2nd edn. Academic Press, Orlando, Florida.

Sharma, A. (1996). Studies of cement dust and heavy metal impact on plant systems around cement industries. Ph. D. Thesis, Vikram University, Ujjain, M.P. India.

Wolff, G.T. and Korsog, P.E. (1989). 'Atmospheric Concentrations and Regional source Apportionments of Sulphates, Nitrates and Sulphur Dioxide in the Bershire Mountains in Western Massachusetts', *Atmos. Environ.* 23: 55-65p.

Chapter 3

Control of Biodeterioration: An Ecofriendly Approach

N.R. Shah[1] & Arun Arya[2]
[1] Deptt. of Museology, Faculty of Fine Arts
[2] Deptt. of Botany, Faculty of Science
M.S. University of Baroda, Vadodara

Abstract

Biodeterioration poses severe problems for preservation of museum materials. There are numerous biological agents, which cause destruction of valuable **cultural property**. Fungus is one of them. **Biodeterioration** can be defined as any undesirable change in the properties of the material of economic importance.

Fungal growth can be seen on a susceptible material if precautionary measures are inadequate. As soon as the presence of fungus is established, it is a usual practice to fumigate or treat the infected objects by using appropriate chemicals. Considering highly toxic nature and carcinogenic effect of various chemicals/ fungicides, it is a high time to find out some natural plant products as a safe and effective alternative for fungicide. In the present investigation an attempt has been made to study some natural plant products for their fungicidal efficacy. Two types of plant products were used for this purpose: 1. Oils and powders and 2. Leaf extracts. Growth of all the four organisms was controlled by 0.5% concentration of sweet flag, Eucalyptus oil was also found effective. Of the various leaf extracts tried margosa and Lantana leaf extracts showed promising results.

Introduction

The control of biodeterioration of cultural property is a very complex affair. Besides preventive measures, chemical, biological and radiological control methods are available. Now a days use of IPM is advocated. Nair (1996) discussing the need for further research in the field of biocides to control the biodeteriogens has stated those vast gaps exists in our understanding on different pests. Effective lethal doses are not known, effects of chemicals on museum objects is not fully understood, the retention effect is not known and effect of harmful chemicals on environment and human health is yet to be worked out.

Use of chemicals may leads to development of resistance in biodeteriogens. At the same time it is feared that if the synthetic chemicals in form of pesticides and food preservatives continue to be used at the present rate, the immunity system of the living beings will weaken and result in complex problems (Bisht, 1991).

Several of the synthetic fungicides have been found to display side effects in form of carcinogenicity, teratogenicity and pollutive effects. Use of biodegradable, less harmful and in true sense ecofriendly natural products of plant origin is slowly replacing the routine fungicides to control plant pathogens (Fawcett and Spencer 1969, Appleton and Tansey 1975, Dixit *et al.* 1978, 1983, Arya 1988 a, b, Arya and Mathew 1990, Arya *et al.* 1995). Attempts are now being made to find safe alternatives to Chlorines that are now either banned in some countries or are treated with suspicion. Chemicals like pentachlorophenol; ethylene dioxide and Dieldrin are very toxic and are hazardous to human beings though effective in killing pests. Research is being undertaken to find effective and chemicals with low toxicity to human body. The neem tree is one such example, its seeds and leaves have a great potential as safe insecticide. Pyrethrum is another insecticide obtained from *Chrysanthemum*. According to Bisht (1991), Panos, a London based information agency, the scientist in the USA are developing biopesticides as ecologically safe alternative to toxic chemicals. They have found that oil extracted from orange and lemon peels is effective against insects but is not harmful to humans.

Museum articles, that are organic in nature, are susceptible to biodeterioration by a wide range of organisms. Materials of plant origin are cellulosic, consisting of carbohydrates and those of animal origin contain proteins. These provide nutrients to the biological agencies of deterioration. Kowalik *et al.* (1962) isolated 87 fungi producing black, brown and yellow pigmentation on paper (Dhawan and Agrawal, 1986). To avoid the irreparable damage by microbes use of preventive measures and control strategies have been suggested by many workers from time to time. Agrawal (1991) points out that attention has not been paid so far, on the effects of fungicides on the materials of cultural property. In the present investigation effect of botanical pesticides was observed *in vitro*.

Materials and Methods

First the infected objects were cleaned with ethanol and when fungal growth reappeared, it was isolated aseptically in a culture tube containing PDA (Potato, Dextrose, and Agar medium) slant. Culture tubes were incubated at 25± 2° C, temperature for eight days. With the help of microscopic examination and morphological characters fungi were identified. They were confirmed by IMI (International Mycological Institute), England. These organisms were used for control experiments.

To observe the effect of oils on the growth of the organisms they were mixed with the PDA medium, while Godavaj powder, in desired concentration was directly mixed after autoclaving. To observe the effect of aqueous leaf extracts, they were homogenized, mixed with distilled water, centrifuged and then diluted to make desired concentrations. These concentrations were used in control experiments. After mixing these plant products in the medium, organisms were inoculated and the growth was recorded after seven days at 25±2° C temperature. Results thus obtained are recorded in Tables 1 and 2.

Oils and powder tried

Cashew nut oil obtained from shells of *Anacardium occidentale* L., Neem oil obtained from kernels of *Azadirachta indica* A. Juiss, Eucalyptus oil obtained from leaves of *Eucalyptus occidentalis* Skeels and Godavaj powder obtained from rhizomes of *Acorus calamus* L. were tried.

Table 1 : Effect of different plant products on percentage control of four biodeteriogens

		Percentage control			
Plant Products	*Conc.%*	*C.verruculosus*	*P.citrinum*	*A.ustus*	*C.cymbiforme*
Cashew nut oil	1	27.2	40	30	36.3
	2	36.3	50	30	36.3
	4	45.4	50	50	36.3
	10	63.6	60	50	45.4
Neem oil	2	16.6	9	22.2	22.2
	5	33.3	18	22.2	22.2
Eucalyptus oil	0.2	46	45.4	36.3	20
	0.5	62.5	63.6	63.6	50
Godavaj powder	0.2	36.3	41.6	66.6	77.7
	0.3	63.6	58.3	77.7	88.8
	0.5	100	100	100	100

Results based on three replicates.

Table 2 : Effect of different leaf extracts on percentage control of four biodeteriogens

		Percentage control			
Leaf extracts	*Conc.%*	*C.verruculosus*	*P.citrinum*	*A.ustus*	*C.cymbiforme*
Neem	100	40	30	50	40
	50	20	10	25	20
Eucalyptus	100	30	20	25	10
	50	20	0	12.5	0
Lantana	100	33.3	50	55.5	50
	50	22.2	25	44.4	30
Lemon	100	0	0	12.5	0
	50	0	0	0	0
Bel	100	18.1	8.3	45.4	30
	50	18.5	8.3	36.3	10
Phudina	100	0	16.6	36.3	0
	50	0	8.3	18.1	0

Results based on three replicates.

Aqueous leaf extracts tried

Search for antimicrobial substances in the tissues of higher plants has been initiated since long. The aqueous leaf extracts of neem, *Azadirachta indica* A. Juiss, Eucalyptus or Nilgiri, *Eucalyptus occidentalis* Skeels, *Lantana camara* L., lemon, *Citrus medica* L., bel, *Aegle marmelos* Corr. and phudina, *Mentha arvensis* L. were prepared. These extracts have been tried *in vitro* in different concentrations and have shown promising results.

Test organisms

Four different fungi *viz. Curvularia* state of *Cochliobolus verruculosus* (Tsuda and Ueyama) Sivan isolated from flute a musical instrument, *Penicillium citrinum* Thom. from Pavri a musical instrument used in religious ceremonies by Kunkuna tribes of Surat (Gujarat), *Aspergillus ustus* (Bainier) Thom. & Church, causing deterioration of wooden Chinese screen displayed in Baroda Museum and *Chaetomium cymbiforme* Lodha present on Parsi manuscript were selected as test organisms to study the efficacy of different products.

Properties of Natural Plant Products used as Fungicides

Following natural plant products were tried against four biodeteriogens.

Cashew nut oil

It is obtained from *Anacardium occidentale* L. of family Anacardiaceae. The fruit of this plant has a kidney shaped nut. Between the two walls of shells, there is a brown colored oil. It has an extremely blistering effect on the skin. This oil is called cashewnut oil. The chief constituents of this oil are Anacardiac acid ($C_{22} H_{33} O_3$), Orthohydroxy benzoic acid and Cardol ($C_{32} H_{52} O_4$) a phenol. Earlier this oil has been used for the preservation of wooden objects (Shah, 1990).

Neem oil

Use of margosa oil is a common feature by the people of Indian subcontinent. The leaves are frequently used as an insect repellent for preserving books, textile and grains. It is extracted from the seed kernels. It has two crystalline bitter principals- Nimbin and Nimbinin, and an amorphous bitter compound Nimbidine.

Eucalyptus oil

Eucalyptus occidentalis Skeels, a member of family Myrtaceae yields volatile eucalyptus oil. The oil is present in the leaves which imparts an aromatic smell. Volatile oil contains Ceniole.

Godavaj powder

It is commercially obtained from *Acorus calamus* L. of the family Araceae. Dry rhizomes of this perennial herb have a mellow odour and contain 1.5 to 3.5 % of yellow, aromatic, volatile oil. It is likely that this oil may contain certain substances with insecticidal properties (Agrawal, 1984). It is a general practice to use Godavaj powder as a repellent in libraries and archives. It is kept in small cotton bags among books and manuscripts in the Oriental Institute of the M.S. University of Baroda.

Leaf extract of Neem

Use of neem leaves is very common as pesticides. In the present study an attempt has been made to study the effect of leaf extract as fungicide. Pandey *et al.* (1983) reported its antifungal effect against a fruit rot pathogen *Pestalotia psidii*.

Leaf extract of Eucalyptus

As leaves of *Eucalyptus* contain essential oil, it was thought desirable to evaluate the effect of leaf extract of *Eucalyptus* also as a fungicide.

Leaf extract of Lantana

It is a member of Verbenaceae. Agrochemicals, Nandesari, Baroda is manufacturing a fungicide named Ovis from the leaves of *Lantana camara*. Considering its potential as a protectant, the leaf extract of *Lantana* was tried *in vitro*.

Leaf extract of Lemon

Lemon (*Citrus medica* L.) is an evergreen shrub or small tree belonging to the family Rutaceae. Its leaves contain volatile oils like neroli and petitgrain. Minimum inhibitory concentration for two fungal deteriogens, *Aspergillus flavus* and *A. versicolor* was 2000 ppm (Dubey *et al.*, 1982).

Leaf extract of Phudina

Phudina (*Mentha arvensis* L.) is a member of family Lamiaecae. The leaves of this plant contain a volatile oil with menthol. This may have an antifungal property. Leaf extract of *M arvensis* var. *piperascens* has been found effective against *Helminthosporium oryzae* a pathogen of paddy (Dikshit *et al.*, 1979).

Leaf extract of Bel

Bel (*Aegle marmelos* Corr.) belongs to the family Rutaceae. The leaves of this plant contain an alkaloid Aegelinine and an essential oil Cineole having antifungal property. The 3000 ppm concentration of this oil was inhibitory to *Rhizoctonia solani* (Renu, 1981).

Results and Discussion

It is evident from Table 1 and Fig. 1 that 10% concentration of cashewnut oil could control 63%, 60%, 50% and 45% growth of *C. verruculosus, P. citrinum, A. ustus* and *C. cymbiforme* respectively. Present study revealed that 5% neem oil could control 33% growth of *C. verruculosus*, 18% growth of *P. citrinum* and 22% growth of *A. ustus* and *C. cymbiforme*. Eucalyptus oil showed more than 60% control of *C. verruculosus, P. citrinum* and *A .ustus* (Fig.1) at 0.5% concentration, whereas, it could control only 50% growth in case of *C. cymbiforme*. In case of Godavaj powder, 0.2% concentration gave 36.3% control of *C. verruculosus*, 41.6% of *P. citrinum*, 66.6% of *A. ustus* and 77.7% of *C. cymbiforme*. This compound completely checked the growth of all the four organisms at 0.5% concentration.

These results clearly indicate that though first three products could not control 100% growth of these organisms at the concentrations tried, yet they may be tried at higher concentrations to achieve the desired results. It was proved that all these products have antifungal properties. Among these Godavaj powder was proved lethal to all the four organisms. Thus, further studies should be conducted on its effect on museum materials. In the International Rice Research Institute, Manila, Philippines, extracts from powdered leaves and seeds have been tested against a variety of pests. Different derivatives of neem have been found to repel 123 species of insects of stored grain (Bisht, 1991). Use of neem oil in conservation of museum materials has been suggested by Dutta (1985-87). Later on Sharma (1992) recommended use of neem oil along with other plants as modern herbal pesticide. Dry rhizomes of sweet flag contain 1.5-3.5 % yellow aromatic volatile oil. It is likely that the insecticidal properties are due to the presence of this oil (Agrawal, 1984).

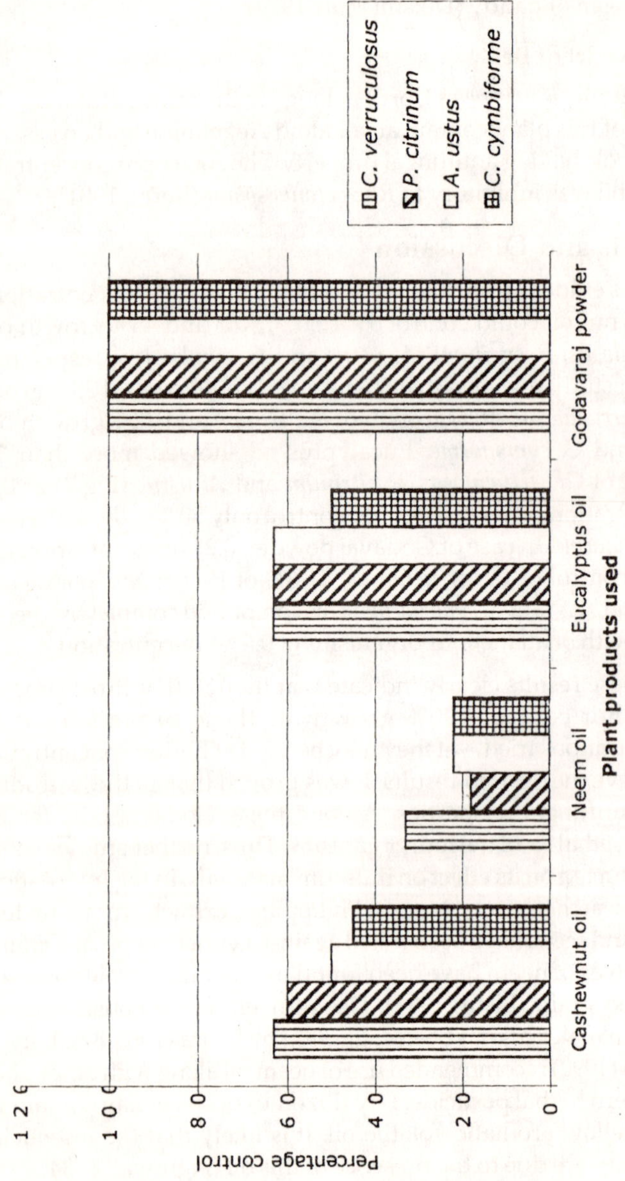

Fig. 1: Effect of plant products on four biodetereogens

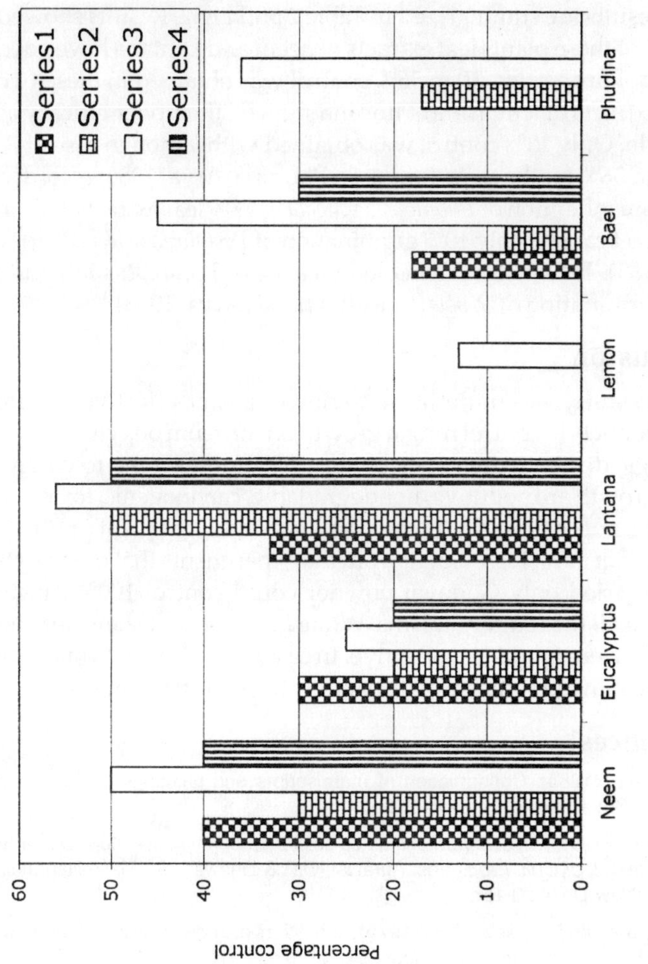

Fig. 2: Effect of leaf extract on four biodetereogens

A number of plants have been reported to possess antifungal substances in their leaves. A study has been conducted to find out the effect of leaf extracts of certain plants on four fungi. Aqueous leaf extracts of neem, *Eucalyptus, Lantana,* lemon, Bel and Phudina were tried. Results are summarized in Table 2 and Fig.2. Results showed that out of these plants leaf extracts of neem and Lantana have good fungicidal properties. 30 to 56% control was obtained in these two cases. Extracts of lemon and Phudina showed their poor efficacy as fungicide. Only 13% control was obtained with lemon in case of *A. ustus* and 36% for the same fungus in Phudina. Arya (1988) recorded 82% spore inhibition of *Phomopsis viticola* in 75% leaf extract of neem this extract caused only 10% germination of *Pestalotia psidii* (Pandey *et al.,* 1983). Presence of Ceneole in leaves of *Eucalyptus* inhibited spore germination of *P. psidii* and *P. viticola* (Arya, 1988).

Conclusion

This study was undertaken to find out some effective natural plant products to control fungal growth on museum objects. Most of the fungicides and chemicals being used in museums to control fungal growth are pollutive, nondegradable, carcinogenic, toxic and therefore, not ecofriendly. In the present study, eight plant products were tried *in vitro*. It is evident from the experiments that out of four products tried only Godavaj powder could control 100% fungal growth in all the four organisms. While *Eucalyptus,* cashew nut and neem oils were found less effective. In cases of leaf extracts majority of plants showed some inhibitory effect to all the test fungi.

References

Agrawal, O.P. (1984). Conservation of manuscripts and paintings of Southeast Asia. Butterworths, London. 299p.

Agrawal, O.P. (1991). An overview of studies on biodeterioration of cultural property. In, *Biodet. of Cult. Prop.* (eds. Agrawal, O.P. & Dhawan, S.) Macmillan India Ltd., New Delhi ; 3-15p.

Appleton, J.A. and Tansey, M.R. (1975). Inhibition of growth of zoopathogenic fungi by garlic extract. *Mycologia* 67: 43-52p.

Arya, A. (1988 a). Control of Phomopsis fruit rots by leaf extracts of certain medicinal plants. In *Indigenous Medicinal Plants Symp.* (ed. Kaushik, P.) Today & Tomorrow's Printers & Publishers, New Delhi: 41-46p.

Arya, A. (1988 b). Control of Phomopsis fruit rots of grapes and guava. *Indian Phytopath.* 41 (2): 214-219p.

Arya, A., Chauhan, R. and Arya, C. (1995). Effect of Allicin and extracts of garlic and *Bignonia* on two fungi. *Ind J. Mycol. Plant Pathol.* 25: 316-318p.

Arya, A. and Mathew, D. (1990). Control of chiku fruit rot by leaf extracts of certain medicinal plants. *Res. J. Plant Env.* 6 (I): 31-33p.

Bisht, A.S. (1991). Control of biodeterioration in museums- A study of problems. In: *Biodet. of Cult. Prop.* (eds. Agrawal, O.P. & Dhawan, S.) Macmillan India Ltd., New Delhi 98-103p.

Dhawan, S. and Agrawal, O.P. (1986). Fungal flora of miniature paintings & lithographs. *Int. Biodet. Bull.* 22 (2): 95-99p.

Dikshit, A., Singh, A.K., Tripathi, R.D. and Dixit, S.N. (1979). Fungitoxic and phytotoxic studies of some essential oils. *Bioi. Bull. (India)* 1: 45-51p.

Dixit, S.N., Srivastava, H.S. and Tripathi, R.D. (1978). *Proc. Indian Soc. Pl. Phisiol.* 80p.

Dixit, S.N., Dubey, N.K., and Tripathi, N.N. (1983). Fungitoxic essential oils vis-a-vis disease control. In *Recent Advances in Plant Pathology* (eds Husain, A., Singh, K., Singh, B.P. and Agnihotri, V.P.). Print House India, Lucknow, 248-252p.

Dubey, N.K., Kishore, N., Tripathi, N.N., Tripathi, R.D., and Dixit, S.N. (1982). *Ann. Appl. Biol.* 100 (supplement 3): 58-59p.

Dutta, P.K. (1985-87). Use of neem oils in conservation. *Con. of Cul. Prop. In India,* New Delhi, 28-30: 98-100p.

Fawcett, C.H. and Spencer, D.M. (1969). Natural antifungal compounds. In *Fungicides II* (ed. Torgeson, D.C.) Academic Press, London, 637p.

Kowalik, R., Sadurska, I. and Czerwinska, E. (1962). Microbiological deterioration of old books remedies. *Bolletino, dell' Instito di Patologia del' Libra alfonso Gallo*, 21: 116-151p.

Nair, S.M. (1996). Biodeterioration: gaps in our knowledge and need for further research. In: *Recent trends in conservation of art heritage.* (O.P. Agrawal's Felicitation volume ed. Dhawan, S.) Agam Kala Prakashan, New Delhi, 181-188p.

Pandey, R.S., Bhargava, S.N., Shukla. D.N. and Dwivedi, D.K. (1983). Control of Pestalotia fruit rot of guava by leaf extracts of two medicinal plants. *Revista Mexicana De Fitopatologia* 2: 15-16p.

Renu (1981). Evaluation of higher plants for their fungitoxicity against *Rhizoctonia solani* Ph.D. Thesis, Gorakhpur University, India.

Shah, A.R. (1990). Application of cashewnut oil as wood preservative. *Con. of Cul. Prop. in India*, New Delhi,23 : 28-29p.

Sharma, B.R.N. (1992). Pest and pesticides in Archaeology and Archives. *Con. of Cul. Prop. in India*, New Delhi, 25: 52-55p.

CHAPTER 4

Green Belt, Plant Scavengers for Combating Air Pollution

Leena A., Jasrai Y.T. and Garge S.K.
Deptt. of Botany, Faculty of Science,
M.S. University of Baroda, Vadodara

Introduction

The industrial revolution gave us all the comforts of modern-day living. But due to reckless industrialization over years, tody we have on our hands something we never bargained for like degradation of our land, contamination of air and water. More over, disappearing forests, depletion of soil cover, floods and droughts, increasing desertification and above all, rising lvel of pollution, are all outcomes of directionless development that we see all arond us.

Pollutant can be regarded as any solid, liquid or gaseous substance present in such a concentration as, may be or tend to be injurious to the human beings or other living creatures or vegetation or property The Air (Prevention and Control) Act, 1981; C_1 (b). With the increasing levels of primary pollutants the concentration of secondary pollutants, particularly of ozone is increasing.

Thus, with the widespread damge it becomes necessary to control the emissions and prevent the environment from getting more polluted. Vegetation can act as a natural sink for most of the atmospheric pollutants which are acting as a Chief living component for bio-geochemical. Cycling in the terrestrial ecosystem. Tree plantations or green-belts are suggested as one of the strategies for industrial air pollution amelioration. Thus, cultivation, conservation or certain trees which are fast growing, resistant or sensitive to certain pollutant can make environment clean to great extent.

Types of Pollutants

Pollutant emitted from different sources can be classified into two types:

1. Primary Pollutants (Tebbens, 1968)

They are emissions directly evolved from an identifiable source of pollutant. e.g. particulates, hydrocarbons, SO_2, nitrogen oxides, carbon monoxid etc.

2. Secondary Pollutants (Fontan & Lopex, 1984)

They are formed by the reaction of two or more emitted or pre-existing chemical in the atmosphere or by the reaction of chemicals with temperature, light etc.

Ozone and perpoxyacotyl nitrate (PAN) are important secondary pollutant emitted by automobile exhaust. Acid rain, smog and radiaoactive substances are secondary pollutants.

Plants as Bio-Sinks

The impact of pollutants on vegetation is much apparent due to their stationary nature. During photosynthesis and respiration, pollutants are absorbed along with CO_2 and O_2. This uptake varies from plant to plant and those having tolerance act as efficient bio-sinks. Pollutants uptake can often be beneficial (Kosta-Rick &* Manning, 1993), or harmful (Friend & Tonlinson, 1992).

There are several factors which influence the pollutant uptake by plants. They are as follows:

1. Growth conditions with respect to light, humdity, water status, nutrients present in soils, day-light period, temperature etc.
2. Concentration of pollutant.
3. Length of exposure - plants can act as more efficient scavengers when subjected to intermitent supply to pollutant rather then continuous supply.
4. Meterological factors affecting the deposition of air pollutant are dew, wind speed and wind direction.
5. Species of plant.

For examples: NO_2 absorption depends on species, Populus have highest rate of NO_2 absorption but is very vulnerable when tested for a mixture of NO_2 and O_3.

6. Resistance of species (Erisman, 1992).

Entry of Pollutant

Once deposited on the foliar surface the entry of pollutant into the leaves occurs either via stomata or by permeating through cultivar layer.

Plants for Scavenging SO_2

SO_2 is the main sulphur compound emitted into the atmosphere. It can be oxidized to SO_3 which forms H_2SO_4 when hydrated.

$$SO_2 \longrightarrow SO_3$$

$$SO_3 + H_2O \longrightarrow H_2SO_4 \longrightarrow \text{aerosol}$$

It is estimate that sulfite is 30 times more toxic than sulfate. So the degree of oxidation of SO_2 to sulfate in the atmosphere has important consequences as far as the plant is concerned (Thomas *et al.*, 1943).

Sources of SO_2

88.0 per cent fuel combustion
8.7 per cent industrial process
3.3 per cent transportation

Plants can detoxify SO_2 to an extent and store it in the form of sulphate (Renneberg, 1984). Large quantities are absorbed by external plant surfaces which acts as filters screening out much of SO_2 in lower atmosphere (Treshow, 1970). Concentration of SO_2 in the range of 0.1 to 0.5 ppm can stimulate stomatal opening (Unsworth *et al.*, 1972). This is especially true with water stressed plants were stomata tended to be partially closed. They reported that SO_2 could even cause partial opening of stomata in dark.

SO_2 due to its higher solubility enter rapidly into the cell. Inside the cell hydrated SO_2 is dissolved into H^+, HSO_3^-, and SO_3^{-2}, when radioactive sulphate is supplied to plant tissue the products include radioactive sulphite, sulphide, cysteine and methionine.

Nutritional status plays an important role in determining damage by SO_2 increased nutrient levels of Ca^{2+} and K^+ tend to lower the damage, while increased phosphate levels increased the damage. Fumigation of sulphur - deficient plant with SO_2, for several weeks showed that SO_2 in the air can be decisive source for sulphur and serves to promote normal plant growth (Faller, 1970).

Species Tolerant to SO_2

Sr. No.	Botanical Name	Common Name	Conclusion
Crop Species*			
1.	*Zea mays*	Maize	2 ppm/8 hrs
2.	*Apium gravelous*	Celery	2 ppm/8 hrs
3.	*Citrus* spp.		2 ppm/8 hrs
Tree Species**			
4.	*Anogeissus latifols*		2.26-4.6 mg/g dry wt.
5.	*Azadirachta indica*	Neem	4.24-12-14 mg/g dry wt.
6.	*Bauhinia racemosa*		2.02-4.90 mg/g dry wt.

** (Saha D, 1998)

Plants for Scavenging Nitrogen Oxides

There are several oxides of nitrogen which may be found in atmosphere, the most important as air pollutants are nitric oxide (NO) and nitrogen dioxide (NO_2)

$$N_2 + O_2 \longrightarrow 2NO$$

$$2NO + SO_2 \longrightarrow 2NO_2$$

Sources of Nitrogen Oxides

50.0 per cent fuel combustion

44.9 per cent transportation

3.9 per cent industrial process

1.6 per cent miscellaneous

There is however, little nitrogen in fuel and nitric oxide is formed in the heat combustion when atmospheric nitrogen and oxygen combines.

When NO and NO_2 dissolve in the extra cellular water within the leaf they form nitrate and nitric ions. They are normally present within a plant cell as a part of the nitrate assimilation pathway in which nitrate is reduced first to nitrite and then to ammonium ions which are used in formation of amino acids and finally proteins.

Exposure to nitrogen increased the activity of enzymes such as glutamate dehydrogenase (involved in reductive amination), glutamate oxaloacetate transaminase and glutamate pyruvate transaminase. As these enzymes are the one involved in the pathway by which nitrogen is assimilated into amino acids, it can be seen that after fumigation with NO_2 there is an increased ability to assimilate nitrogen.

Species Tolerant to Nitrogen Oxides

Quantities of nitrogen removed by various crops (only the nitrogen found in parts normally harvested earlier)

Sl. No.	*Crop*	*Nitrogen Removed (kg/h)*
1.	Tomato (fruit)	130
2.	Maize (grain)	150
3.	Barley	40
4.	Wheat	50
5.	Sugarcane	110
6.	Grapes	110

Tree species and ornamental spp. tolerant to nitrogen oxides are:

Sl. No.	*Tree*	*Concentration*
1.	*Ginkgo biloba*	10 ppm/4-8 hrs.
2.	*Quercus penduncculata*	"
3.	*Taxus baccata*	"
4.	*Pinus austriaca*	"
Ornamental Spp.		
1.	*Carissa carandas*	15 ppm/4 hrs.
2.	*Codiarm variegatum*	"
3.	*Juniperus coniferatas*	"

Plants for Scavenging Ozone

Minor amounts of ozone are added to atmosphere by electrical discharge such as lighting flashes, more may be brought down from upper atmosphere by vertical flux. Inefficient internal combustion engines daily emit tons of waste hydrocarbons and nitrogen oxides into the atmosphere.

$$NO_2 \xrightarrow{\text{light}} NO + O$$

$$O + O_2 \longrightarrow O_3$$

Ozone moves passively down a concentration gradient much CO_2 diffusion during active photosynthesis. The removal rates for ozone are greatest to lush vegetation and less to, water-stressed vegetation. The entry of this pollutant escalates the formation of free radicals in plants cell. To counter act the damaging effects of free radicals, plants possess natural scavengers which remove free radicals at varoius steps to minimize the damage. Ascorbate and glutathione are such anti-oxidants.

Tolerant Species

53-56 ppm/4 hrs.	Cotton
Beet	Lettuce
Geranium	Bur oak
Cucumber	Siberian elm

Plants for Scavening Carbon Monoxide

Carbon monoxide is a trace constituent of the troposphere produced by both natural processes as well as human activities.

Sources of CO

77.4 per cent transportation

7.2 per cent industrial process

5.6 per cent fuel combustion

9.8 per cent miscellaneous

Because plants can both metabolize and produce CO in the vicinity of urban industrial area can substantially exceed global background levels. There are no reports of these currently measured levels of CO producing any adverse effects on plants. CO incomplete burning of natural gas, kerosene etc. is also an common indoor air contaminant. Genated by highly toxic levels of CO were reduced to non-toxic levels in 24 hrs by a single spider plant (Cloropytum comosum) in a small encolsed space. 96 per cent of the CO was removed by this plant (Dr. Bill Wolverton, NASA research, 1985).

Top 10 plants most effective in removing CO are:

1.	Bamboo palm	:	*Chamaedorea seifritzii*
2.	Chinese evergreen	:	*Aglkaonema modestrum*
3.	English ivy	:	*Hedera helix*
4.	Gerbera Daisy	:	*Gerbera jamesoni*
5.	Janet Craig	:	*Dracaena* spp.
6.	Marginata	:	*Dracaena marginata*
7.	Mass cane	:	*Dracaena massangcane*
8.	Mother-in-law tongue	:	*Sansevieria laurentii*
9.	Pot mum	:	*Chrysanthemum morifolium*
10.	Peace lily	:	*Sapathiphyllum* spp.

Plants for Scavenging Indoor Pollutant

People worry so much about outdoor air, but indoor may be far more serious problem said Dr. Wolverton. Most common indoor pollutants are formaldehyde, benzene and trichloloethylene released mainly by burning natural gas, kerosene and cigarette smoking.

Sl. No.	*Toxin*	*Plants that filter it out*
1.	Benzene	*Gerbera jamesomi* *Chrysanthemum morifolium, Dracaena marginata, Hedera helix.*
2.	Formaldehyde	*Chlorophytum comosum, Chamaedorea seifritzii, Philodendron*
3.	Trichloroetlylene	*Chrysanthemum, Gerbera jamesomi, Spathiphyllum, Dracaena deremensis.*

Green Belt

Tree plantations around any industrial area forms the "green belt" Forests are known to take up atmospheric pollutants (Van stempvoult *et al.*, 1991). Green belt plantation around air polluting units can never be a claim for the removal of air pollutants at the region. Yet, effectively planted trees in a green belt can potentially remove considerable amounts. Tree plantations or green belts therefore are suggested as one of the strategies for industrial air pollutants amelioration.

Pollution dilution over a broader area is a commonly used method which need modification at source level. Transport of pollutants from source to the forest (receptor) begins with turbulent diffusion into the atmosphere at the source during which a substantial transformation of chemical form can take place. Transport is accomplished by air parcels (eddies) of higher concentration downward to the forest. This indicates that checking air pollution near the source is very important.

Studies indicate the green belt plantation should be based on stack height, wind flow (metrology), types of pollutants etc. Based on two types of stack height, two separate plantation strategies can be adopted.

1. For High Level Sources

The deposition near the source is less and increases with the down-wind until it reaches maximum and decreases again. As there will be less deposition in the wake region, plantation of moderate and sensitive species can be mixed near the source with an increase in concentration down-wind, upto it's maximum, there could be gradual increase in the number of tolerant species in combination with moderate ones. At the point of maximum concentration/ deposition, only tolerant species should be planted. Again in the periphery a combination of moderate and sensitive ones can be planted.

2. For Ground Level Sources

The dry deposition takes place near to the source where concentrations are highest and decreasing down-wind (Erismaun, 1992). So near the ground level sources only tolerant species should be planted. They can be surrounded by a mixture of tolerant and moderate tree.

Green belts for both the patterns of pollutant settlement usually should be in circular-form around sources. For greater efficiency, circular green belts can be winded in prevailing wind flow direction. It is advisable to keep fast growing tolerant species in close vicinity to the source and thorny species out side the green belt for protection.

Literature shows that seedling and sapling of trees are more sensitive to air pollution. Hence plantations should be done well before the actual functioning. This will give an established and functional green belt. Thus, green belt should not be developed only as an air purifier but it should be allowed to develop as a major component of the industrial ecosystem by increasing number of species and their interaction.

Conclusion

The studies have revealed that plants do play a major role in removing or reducing the pollution levels. Such plants would be of great help for creating/developing green belts, which should be a major component of industrial ecosystem. And also, for selection of crop species for agricultural land surrounding the industries. Also, genetic studies can help to develop more new resistant varieties. Some plants are so efficient in absorbing indoor contaminants in the air, that some will be launched in space as a part of biological life support.

References

Allen, S. Heangel *et al.* (1991). *New Phytology.*

Guderian, R. (1977). *Air Pollution.* Springer Verlag, Germany.

Hill, M.K. *Understanding Environmental Pollution.*

Mudd, J.B. & Kozlowski (1975). *Responses of Plants to Air Pollution.* Academic Press, London.

Nebel, B.J. & Wright, R.T. (1996). *Environmental Science.* Prentice Hall, New Jersey.

Treshow, M. (1970). *Environment and Plant Response.* McGraw Hill Book Company.

Saha, D. (1998). *Tolerance Levels of Certain Tree Species Towards Industrial Air Pollution,* Ph.D. Thesis, M.S. University of Baroda.

Unsworth, M.H. & Ormrod, D.P. (1982). *Effects of Gaseous Air Pollution in Agriculture and Horticulture.* University Press, Cambridge.

CHAPTER 5

A Study on Comparative Response of a Crop and Weeds Associated with it, Against Sulphur Dioxide Toxicity

D.M. Kumawat, P. Pandya and N.K. Jain

Institute of Environment Management and Plant Sciences
Vikram University, Ujjain (M.P.)

Abstract

Weeds are more adaptive and competitive in nature than most of the crops for withstanding various physical and edaphic stresses. Literature regarding their adaptive behaviour against atmospheric pollution is scarce. In the present study an attempt was made to acquire knowledge and information regarding the behaviour of certain weeds against sulphur dioxide stress. Five weeds commonly growing in association with winter wheat crop (*Triticum aestivum* L. cv. WH-147), namely *Avena fatua* L., *Setaria glauca* L., *Phalaris minor* Retz., *Chenopodium album* L. and *Anagallis arvensis* L. were selected for the present study. The crop and weed plants were subjected to sulphur dioxide (SO_2) at a concentration 1310 $\mu g/m^3$ in agricultural fields at Nagjhiri (Ujjain). Various physiological and biochemical parameters of both, crop and weed plants for their respective response against SO_2 stress were evaluated. Comparatively the three grassy weeds *A. fatua, S. glauca* and *P. minor* were significantly less affected than the crop plants (wheat). Interestingly the broad leafed weeds *A. arvensis* and *C. album* revealed a reverse response and were more affected than the crop plants.

Introduction

Agriculture is and for years to come will remain India's greatest sort of industry and the foundation of the state (Howard, 1982). Agriculture as our main source of survival, so with limited cultivable land area, fluctuating climatic and weather conditions and above all a huge population to feed, we cannot afford to forego the consideration for agricultural betterment. Weeds are the main menace in agricultural fields throughout the world. Weeds compete with crop plants in sharing nutrients, soil moisture, sunlight and space. Weeds are an important factor in the management of all land and water resources, but their effect is greatest on agricultural yields. Reduction in crop yield has a direct correlation with weed output. Generally, an increase of one kilogram of weed growth corresponds to reduction of one kilogram of the crop growth.

Sulphur dioxide has been recognised as one of the important gaseous pollutant causing severe crop loss throughout the world. Weeds and crop plants both are subjected to same pollution stress in the fields. Do the weeds have certain adaptations to scavenge or to resist such atmospheric stresses? What is their response against such stresses? Do they still pose the same competition for the various aforesaid factors? With these and many more queries, in the present study an attempt was made to evaluate the comparative response of few common weeds growing along with one of the important cereal crop i.e., wheat.

Materials and Methods

Plant selection and SO_2 treatment

After a survey, agricultural fields with severe weed infestation were selected at Nagjhiri area in Uijain (Madhya Pradesh) for the present study. The study was carried out during winter season and the crop under cultivation was wheat (*Triticum aestivum* L. cv. WH-147). Five dominant weeds growing associated with the crop namely *Anagallis arvensis* L. *Avena sativa* L., *Chenopodium album* L., *Phalaris minor* Retz. and *Setaria glauca* L. were selected for the study. Following Sharma and Thakre (1980) the weeds and the crop plants were subjected to SO_2 stress at a concentration of 1310 μm^3 for four hours daily (7-9 a.m. and 4-6 p.m.) for one month. The SO_2 treatment was provided 25 days after crop emergence in the fields. The treatment was carried out in polythene chambers (lx1x1 m^3) and the SO_2

concentration was monitored time to time using Toxic Gas Monitor-555 (CEA., USA).

Plant Assay

Various physiological and biochemical parameters were evaluated of both the crop and weed plants to assess their response following standard methods proposed by different workers. Buffering capacity, total foliar protein content, lipid peroxidation study, free proline content, total water extractable sulphydryl (-SH) content, Nitrate reductive activity and total foliar sulphate content was estimated in all the plants after one month of exposure to SO_2, following Scholz and Reck (1976), Lowry *et al.*, (1951), Dhindsa and Matowe (1981), Bates *et al.*, (1973), Grill *et al.*, (1979), Srivastava and Mathur (1980) and Patterson (1958) respectively. The stomatal conductance study was made using Steady State Porometer (Li Cor 1600, USA).

Statistical Analysis

To test the existing significance between the means of treatment, Student's t-test was employed. Any value less than 5% was rejected.

Results and Discussion

The reaction of plants in response to air pollutants is strongly dependent on the species investigated and might also depend on the physiological state of the plant (Kunert and Hofer, 1987). Since stomata are the chief portals for gaseous exchange in plants, their functioning has been reported to be altered by gaseous pollutants like SO_2, O_3, H_2S etc.

In the present study, under SO_2 stress the stomatal conductance decreased in case of *A. fatua*, *S. glauca*, *P. minor* and *C. album* while it increased in *A. arvensis* and the crop plants (Table 1). The above observation suggests that at least four of the weeds have the adaptation of stress avoidance and thereby resulting in a low uptake of the pollutant.

Only a small part of SO_2 absorbed by the leaves is directly reduced to sulphide and incorporated into organic sulphur, most of it is rapidly converted to and accumulated as sulphate (Filner *et al.*, 1984). The observation made for stomatal conductance further gets support from the sulphate (SO_4^{-2}) content data. The sulphate content increased in all the plant species studied, over their respective control values (Table 2). The increase in SO_4^{-2} content was least in case of *A. arvensis*, followed by *C. album, P. minor, S. glauca* and *A. fatua*. In crop plants there was a significant increase in the SO_4^{-2} content. If plants are not able to metabolise all the SO_2 to non- toxic organic compounds or sulphates, increased toxicity of sulphites can be noticed (Miller and Xerikos, 1979). The sulphite to sulphate conversion rate is faster in tolerant plants, which enables them to escape the toxicity of sulphite ions that is about thirty times toxic than sulphate ions. In the present study the three grassy weeds *Avena, Setaria* and *Phalaris* along with crop plants had more accumulation of SO_4^{-2} suggesting a faster conversion rate in them. In case of dicot weeds *Chenopodium* and *Anagallis* the SO_4^{-2} content was considerably less, may be due to a slower conversion rate of SO_3^{-2} to SO_4^{-2}.

Table 1 : Effect of SO_2 on stomatal conductance (cm/s) in wheat and different weeds associated with it

Plant species	*Control*	*SO_2 treated (1310 µg/m³)*
Triticum aestivum cv. WH-147	0.39 ± 0.03	0.42 ± 0.11
Avena fatua	0.32 ± 0.04	0.27 ± 0.02
Anagallis arvensis	1.22 ± 0.09	1.28 ± 0.10
Chenopodium album	1.18 ± 0.14	1.08 ± 0.04
Phalaris minor	0.37 ± 0.09	0.31 ± 0.06
Setaria glauca	0.35 ± 0.04	0.31 ± 0.02

± Standard deviation

Table 2 : Effect of SO_2 on total foliar sulphate content (mg/g dry weight) in wheat and different weeds associated with it

Plant species	*Control*	*SO_2 treated (1310 µg/m³)*
Triticum aestivum cv. WH-147	1.41 ± 0.06	1.65 ± 0.05
Avena fatua	1.59 ± 0.06	1.70 ± 0.08*
Anagallis arvensis	1.43 ± 0.10	1.46 ± 0.12
Chenopodium album	1.81 ± 0.16	1.85 ± 0.10
Phalaris minor	1.39 ± 0.19	1.44 ± 0.12
Setaria glauca	1.26 ± 0.03	1.36 ± 0.05

* Mean value significant at ($p<0.05$)
± Standard deviation

Susceptibility of plants to SO_2 is related to the pH of their leaf sap. SO_2 is an acidic gas and it generates different ions during its oxidation to a stable product, sulphate. Changes in buffering capacity of the fumigated plants suggest that SO_2 might have affected the overall ionic composition of plants. The buffering capacity decreased in all the plants when compared to their respective controls. In case of *A. arvensis* and the crop plants a significant decrease in buffering capacity was observed (Table 3). Similar decrease in buffering capacity has also been reported in SO_2 treated plants by various workers (Keller, 1982; Darrall and lager, 1984, Bytnerowicz *et al.*, 1987).

Nitrate reductase (an important enzyme in the assimilation of exogenous nitrogen, NR, EC. 1.6.6.1) activity, and the total foliar protein content too were evaluated. A general decrease in both the parameters was observed in SO_2 subjected plants over their respective controls (Table 4 and 5). In *A. fatua* however, a slight increase in protein content over control was observed. A significant decrease was observed in both the dicot weeds, *C. album* and *A. arvensis*. Decrease in NR activity and protein content in SO_2 exposed plants have also been reported by several earlier workers.

Proline content has been reported to increase during any kind of stressed condition in the plants. In the present study too an increase in proline content was observed in all the plant species exposed to SO_2 stress. The increase however was significantly more in case of the three grassy weeds *Avena, Setaria* and *Phalaris* and was least in

Table 3: Effect of SO_2 on buffering capacity of wheat and different weeds associated with it

Plant species	*Control*	*SO_2 treated (1310 µg/m³)*
Triticum aestivum cv. WH-147	2.67 ± 0.06	2.04 ± 0.12*
Avena fatua	2.27 ± 0.12	2.17 ± 0.11
Anagallis arvensis	3.13 ± 0.25	2.77 ± 0.15*
Chenopodium album	3.53 ± 0.15	3.47 ± 0.09
Phalaris minor	3.22 ± 0.09	3.14 ± 0.10
Setaria glauca	2.43 ± 0.13	2.34 ± 0.05

* Mean value significant at ($p<0.05$)
± Standard deviation

Table 4: Effect of SO_2 on Nitrate Reductase (NR) activity (μM NO_2/g fr. wt./h) in wheat and different weeds associated with it

Plant species	*Control*	*SO_2 treated (1310 μg/m³)*
Triticum aestivum cv. WH-147	4.20 ± 0.07	3.85 ± 0.15*
Avena fatua	4.27 ± 0.07	4.15 ± 0.22
Anagallis arvensis	2.95 ± 0.23	2.41 ± 0.42*
Chenopodium album	2.68 ± 0.23	2.39 ± 0.35*
Phalaris minor	3.25 ± 0.19	3.14 ± 0.17
Setaria glauca	3.74 ± 0.24	3.55 ± 0.14**

* Mean value significant at ($p<0.05$), ** Mean value significant at ($p<0.01$)
± Standard deviation

Table 5: Effect of SO_2 on total foliar protein content (mg/g dry weight) in wheat and different weeds associated with it

Plant species	*Control*	*SO_2 treated (1310 μg/m³)*
Triticum aestivum cv. WH-147	44.67 ± 1.96	40.47 ± 1.19
Avena fatua	34.47 ± 1.88	34.97 ± 2.13
Anagallis arvensis	46.07 ± 2.72	37.87 ± 2.17*
Chenopodium album	46.13 ± 1.29	40.13 ± 0.86
Phalaris minor	30.24 ± 1.13	29.35 ± 1.26
Setaria glauca	36.53 ± 2.53	34.17 ± 1.58

* Mean value significant at ($p<0.05$)
± Standard deviation

Table 6: Effect of SO_2 on total free proline content (mg/g fresh weight) in wheat and different weeds associated with it

Plant species	*Control*	*SO_2 treated (1310 μg/m³)*
Triticum aestivum cv. WH-147	0.77 ± 0.15	0.97 ± 0.21*
Avena fatua	0.70 ± 0.17	1.33 ± 0.15**
Anagallis arvensis	1.40 ± 0.20	1.50 ± 0.30
Chenopodium album	1.33 ± 0.23	1.57 ± 0.31*
Phalaris minor	0.84 ± 0.20	1.24 ± 0.18**
Setaria glauca	0.77 ± 0.15	1.27 ± 0.12

* Mean value significant at ($p<0.05$), ** Mean value significant at ($p<0.01$)
± Standard deviation

Table 7: Effect of SO_2 on total water extractable sulphydryl (-SH) content (µmoles/g fr. Wt.) in wheat and different weeds associated with it

Plant species	*Control*	*SO_2 treated (1310 µg/m³)*
Triticum aestivum cv. WH-147	0.79 ± 0.03	0.82 ± 0.09
Avena fatua	0.75 ± 0.09	0.93 ± 0.18
Anagallis arvensis	1.33 ± 0.11	1.23 ± 0.24
Chenopodium album	1.49 ± 0.15	1.57 ± 0.15
Phalaris minor	0.81 ± 0.12	0.92 ± 0.18
Setaria glauca	0.90 ± 0.07	1.09 ± 0.13*

* Mean value significant at ($p<0.05$)
± Standard deviation

Table 8: Effect of SO_2 on lipid peroxidation (Malondialdehyde (MDA)) content (µmoles/g dry Wt.) in wheat and its associated weeds

Plant species	*Control*	*SO_2 treated (1310 µg/m³)*
Triticum aestivum cv. WH-147	1.32 ± 0.05	1.22 ± 0.02
Avena fatua	1.37 ± 0.05	1.12 ± 0.15
Anagallis arvensis	1.76 ± 0.08	1.72 ± 0.05
Chenopodium album	1.77 ± 0.11	1.76 ± 0.13
Phalaris minor	1.25 ± 0.10	1.15 ± 0.18
Setaria glauca	1.28 ± 0.12	1.13 ± 0.12

± Standard deviation

case of *A. arvensis* (Table 6). Proline is believed to protect proteins and membranes from free radicals generated by SO_2 stress. Being a compatible solute it could also function as an osmolyte and may serve as storage compound for nitrogen.

The total water extractable -SH (Sulphydryl) compounds also showed an increase over control values in all the plants studied. The increase was considerably more in case of grassy weeds than dicot ones (Table 7). The crop response was more closer to the dicot weeds. Similar reports of increase in -SH content on exposure to SO_2 has been made by various workers. Lipid per-oxidation measured in terms of malondialdehyde (MDA) content showed a decrease in all the plants studied. The decrease however was more in case of the three grassy weeds than the crop and dicot weeds (Table. 8). Free radicals generated during SO_2 oxidation to SO_4^{-2}, may induce, lipid per-oxidation which is an important mechanism of membrane deterioration (Irigoyen *et al.*, 1992; Navari-lzzo *et al.*, 1992).

In general the present study suggests that weeds do have an adaptive potential for atmospheric stress as well. At least for SO_2 stress the weeds showed considerable avoidance/tolerance potential. The grassy weeds i.e., *A. fatua, S. glauca* and *P. minor* were more sturdy in behaviour than the dicot weeds. Further study is needed to know the exact biochemical mechanism for their tolerance.

Acknowledgemets

The authors wish to thank Prof. and Head, I.E.M.P.S., Vikram University, Ujjain for providing necessary laboratory facilities and to the farmers of Nagjhiri, Ujjain for their co-operation.

References

Bates, L.S., Waldren, R.P. and Tears, I.D. (1973). Rapid determination of free proline for water stress studies. *Plant and Cell* 39: 205-207p.

Bytnerowicz, A., Olszyk, D.M., Kats, G., Dawson, P.I., Wolf, I. and Thompson, C.R. (1987). Effects of SO_2 on physiology, elemental content and injury development of winter wheat. *Agric. Ecosystem Environ.* 20: 37-47p.

Darrall, N.M. and Jager, H.J. (1984). Biochemical diagnostic tests for the effect of air pollution on plants. In: *Gaseous Air Pollutants and Plant Metabolism* (Eds. M.J. Koziol and F.R. Whatley) Butterworths, London, 333-349p.

Dhindsa, R.S. and Matowe, W. (1981). Drought tolerance in two mosses correlated with enzymatic defence against lipid peroxidation. *L Exp. Bot.* 32: 79-91p.

Filner, P., Rennenberg, H., Sekiya, I., Bressan, R.A., Wilson, L.G., Le Cureux, L. and Shimei, T. (1984). Biosynthesis and emission of hydrogen sulphide by higher plants. In: *Gaseous Air Pollutants and Plant Metabolism* (Eds. M.J. Koziol and F.R. Whatley) Butterworths, London, pp. 291-312p.

Grill, D., Esterbauer, H. and Klosch, U. (1979). Effect of sulphur dioxide on glutathione in leaves of plants. *Environ. Pollut.* 19: 187-194p.

Howard, A. (1982). In : *Crop Production in India- A Critical Survey of its Problems.* Oxford Univ. Press.

Irigoyen, J.I., Emerich, D.W. and Sanchez-Diaz, M. (1992). Alfalfa leaf senescence induced by drought stress: photosynthesis, hydrogen peroxide metabolism, lipid peroxidation and ethylene evolution. *Physiol. Plant.* 84: 67-72p.

Keller, T. (1982). Physiological bioindications of an effect of air pollution on plants. In : *Monitoring of Air Pollutants by Plants.* (eds. L. Steubing and H.J. lager), Junk, The Hague, 85-95p.

Kunert, K.I. and Hofer, G. (1987). Geben Veranderungen des antioxidativen systems von pflanzen hinweise auf die wirkung von Luftschadstoffen? *Allg Forstz.* 27: 697-699p.

Lowry,O.H., Rosenbrough, N.J., Farr, A.L. and Randall, R.J. (1951). Protein measurement with the folin phenol reagent. *J. Biol. Chem.* 193: 267-275p.

Miller, J.E. and Xerikos, P.B. (1979). Residence time of sulphite in SO_2 "sensitive" and "tolerant": soybean cultivars. *Environ. Pollut.* 18: 259-264.p

Navari-Izzo, F., Quartacci, M.F., Izzo, R. and Pinzino, C. (1992). Degradation of membrane lipid components and antioxidant levels in *Hordeum vulgare* exposed to long term fumigation with SO_2. *Physiol. Plant.* 84 : 73-79p.

Patterson, G.D. Jr. (1958). Sulphur In : *Colorimetric Determination of Non Metals.* (Ed. D.F. Boltz) Inter Science Publ. Inc. New York, pp. 261-308p.

Scholz, P. and Reck, S. (1976). Effect of acid on forest trees as measured by titration in vitro inheritance of buffering capacity in Picea abies. Proc. *First International Symp. on Acid Precipitation and the Forest Ecosystem*, USDA Forest Service General Technical Report NE -23. pp. 975-976p.

Sharma, H.B. and Thakre, R.A. (1980). Generation and monitoring of gaseous pollutants. *Ind. J. Air. Pollut. Contrl.* 5(1): 24-27p.

Srivastava, H.S. and Mathur, S.N. (1980). Nodulation and NR activity in nodules and leaves of black gram (*V. mungo*) as affected by varying day length. *Ind. J. Exp. Biol.* 18: 300-302p.

CHAPTER 6

Vehicular Exhaust Emissions Control in Spark Ignition Engines with Catalytic Converter System

S. Bhagwanta Rao[1], H.B. Mathur[2] & M.K.G. Babu[3]

[1] *Mech. Engg. Deptt., Dayalbagh Educational Institute, Dayalbagh, Agra*

[2] *Mech. Engg. Dept. D.C.E., Delhi*

[3] *Center of Energy Studies, I.I.T., Delhi*

Abstract

With rapid development in industrial, transport and power generation sectors, clean air has become a rare commodity for the people. In spite of several methods adopted to reduce air pollution, the related problems remain unabated in most of the urban areas. The present work highlights the performance of some of the devices adopted for reducing pollution from automobile vehicles. The cost of such devices should not become a hindrance in development of the Country but add to it in order to reduce the menace of pollution for a prosperous secure and healthier society.

Introduction

With the rapid rise in transportation, the problem of having a cleaner environment is becoming increasingly difficult. About 60-70 per cent air pollution is from motor vehicles operating on either gasoline or diesel oil. In average motor vehicle with no emission control devices (1) the estimated values of annual production of the pollutants are about 770 kg of carbon monoxide, 240 kg of unburned hydro-carbons and 40 kg of Nitrogen oxides. The quantities get increased with increase in the number of vehicles, specially the two and three wheelers that emit large quantities of unburned hydrocarbons along with carbon monoxide and Nitrogen Oxide. Hence the need for development of suitable techniques for reducing these pollutants cannot be ruled out. Out of many techniques tried such as—(a) modification of combustion system, (b) induction system, (c) ignition system, (d) exhaust gas re-circulation, (e) after burners, (f) catalytic converters, Catalytic converters being a relatively low temperature devices can easily be located in exhaust tail pipe and requires no modification in the engine design. These converters are designed to convert the harmful exhaust gases into less harmful gases through chemical reduction and oxidation while the gases pass over a bed of active material either deposited-on supports or unsupported.

Various materials as (1) noble metals (2), base metals as Nickel, Copper, Zinc alloy (3) and their oxides have been used as catalyst in the converters to reduce the pollutants. In order to reduce the cost of the replacement, it is felt to use a cheaper material as catalyst preferably indigenous sources which would have same activation period as that of noble metals. In present work Monel (an alloy of Nickel and Copper) and Sponge Iron have been taken as catalysts. Monel as small chips and sponge iron in the form of small pellets, were used as un-supported catalysts and studies were made under a various operating conditions of the engine operated on gasoline. The catalytic converter fitted in the exhaust line has been found to reduce the pollutants by 60-70 per cent under different load conditions at various speeds and mixture ratios. The introduction of secondary air and its heating has considerably increased the conversion efficiency of the pollutants. Although the some power loss has been observed due to the introduction of the converter, it is not very high compared to the reduction in the pollutants. The cost of such device with indigenous catalysts is much lower compared to other devices.

Experimental setup and procedure

A Ford four cylinder spark ignition engine used in passenger cars, was used to conduct the experiments for the present study of evaluation of catalytic converter system. The engine was coupled to a hydraulic dynamo- meter so as to measure the power output. A

Solexdown draft type carburetor with adjustable main jet was fitted to vary mixtures strength that enters the engine.

Preparation of the Catalyst

A catalyst Monel was available in the form of thin sheet. Small chips had been cut from the sheet which were treated with aquaregia so as to have a rough surface for better contact with the gases. The chips were washed with water and dried up in the oven before fillings the converter. Sponge iron was available in the form of small pellets. These contain some moisture. The pellets had been classified into various sizes and dried up to remove the moisture before packing them in the converter.

Catalytic Converter

It was a cylindrical shell opened at both ends. The front end through a reducer was connected to exhaust manifold while the other end was connected to exhaust line of the engine. Provisions were made to measure the temperatures of selective points with thermo couples. A manometer was used to measure the pressure drop across the converter. The experiments were carried out in two phases on automotive spark ignition engine. In first phase, the tests were conducted to evaluate the performance, fuel economy and emission characteristics with no secondary air, while second phase involved the repetition of the tests using secondary air its heating with catalytic converter in exhaust line. Table 1 gives over all picture of test matrix to generate large variation in operating parameter. The over all set up of engine and other measuring instruments has been shown in Fig. 1.

Table 1

*Speed revolution (per minute)**	*1500*	*1750*	*2000**	*2200*	*2500**
Equivalence ratio	0.9	1.0	1.1	1.2	1.3
Throttle opening	0.25	0.5	0.75		
Secondary air (/ of primary air)	0.0*	4.0	8.0		
Temperature (Deg. K)	300*	350	370		

*Denotes the base line conditions.

Results and Discussions

A series of experiments were conducted to assess the effectiveness of the two catalysts under different conditions of

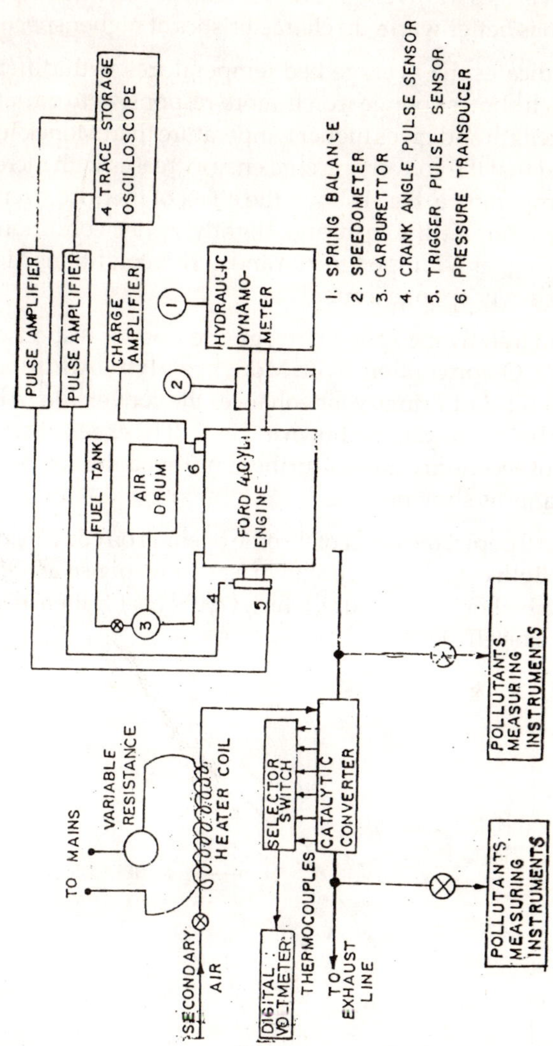

Fig. 1 : Schematic Diagram of Experimental Setup

operation of the engine as mentioned in the Table 1. Graphs were plotted to find out the trend of the converter characteristic with both catalysts. Some typical graphs are shown in Fig. 2 through Fig. 8.

Fig. 2 envisages the average catalyst temperature with speed. Sponge iron has better warm up characteristics at higher speeds.

Fig. 3 indicates the average bed temperatures with different catalysts which shows sponge iron is more responsive to variation in mixture strength and gives higher temperature than Monel. It has been observed that the conversion efficiency increases with increase in catalyst temperature. Fig. 4 shows the effect of mixture strength no No-conversion. At the lean and slightly richer condition no conversion is higher with sponge iron and at rich conditions Monel shows an edge over sponge iron.

Fig. 5 and 6 show the sponge iron giving a better response for the HC and CO-conversion over Monel catalyst. Addition of secondary air 6-8 / of Primary air enhances the conversion of HCs and CO with both catalysts as shown in Fig. 7. It has been observed that heating of secondary air has further improved the conversion of the pollutants as shown.

Fig. 8 and 9 depict the effect of throttle opening on conversion of HC and CO. Both catalysts have shown a good response in getting heated up to a higher temperature, which would enable them to give better conversion efficiency.

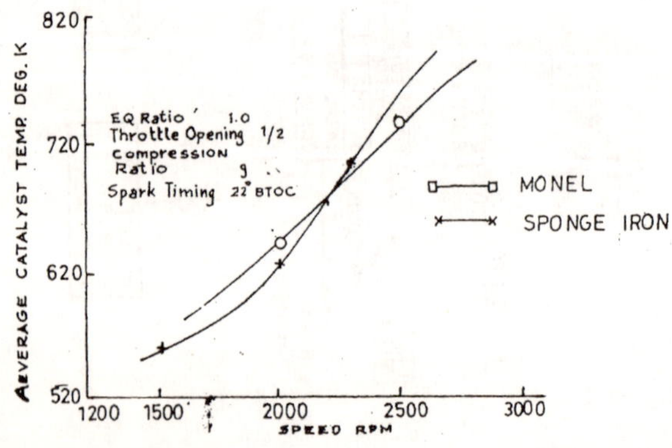

Fig. 2 : Effect of Speed

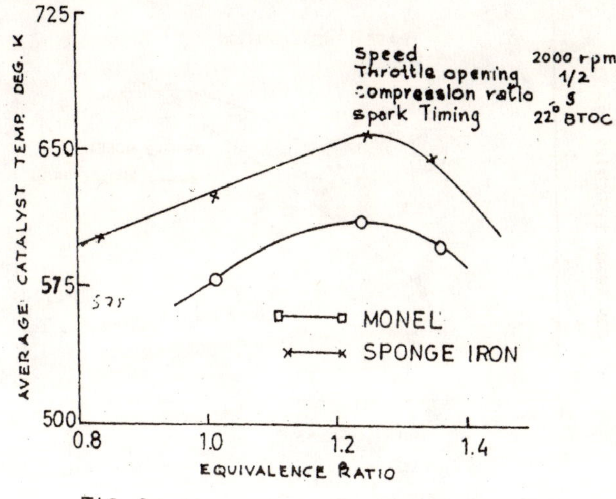

Fig. 3 : Effect of Mixture Strength

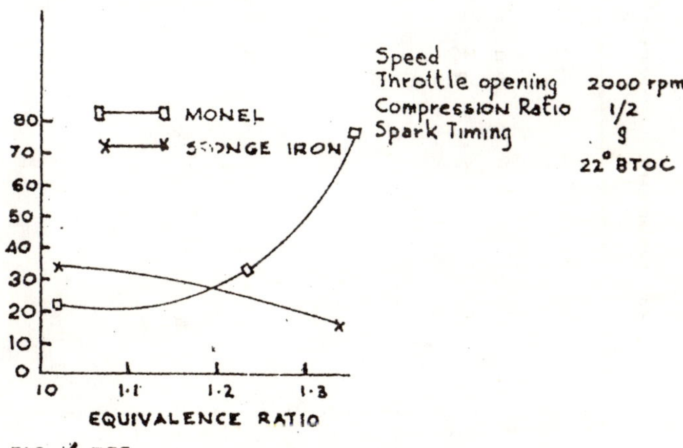

Fig. 4 : Effect of Mixture Strength on No-Conversion

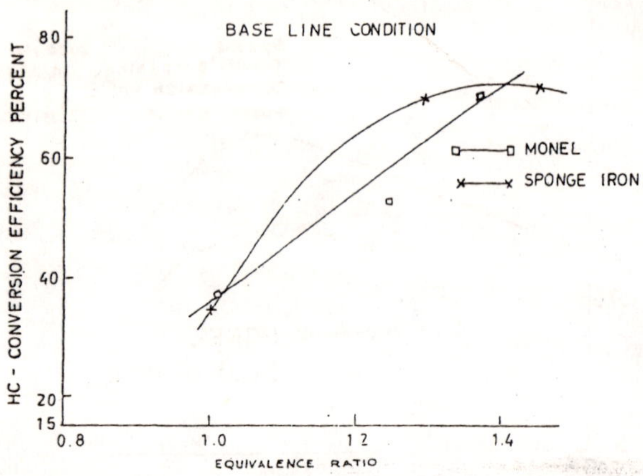

Fig. 5 : Effect of Mixture Strength on HC-Conversion

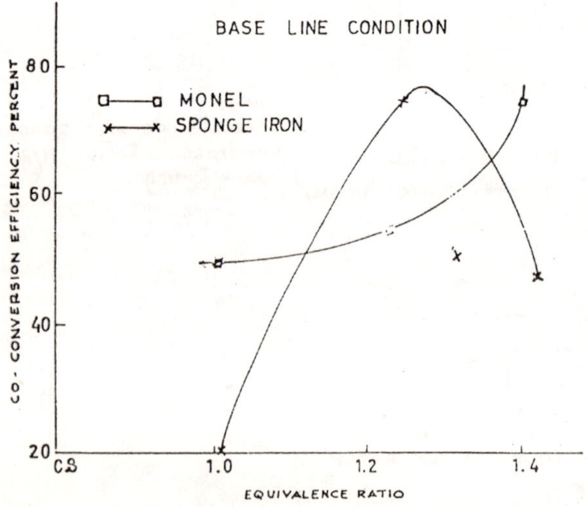

Fig. 6 : Effect of Mixture Strength on CO-Conversion

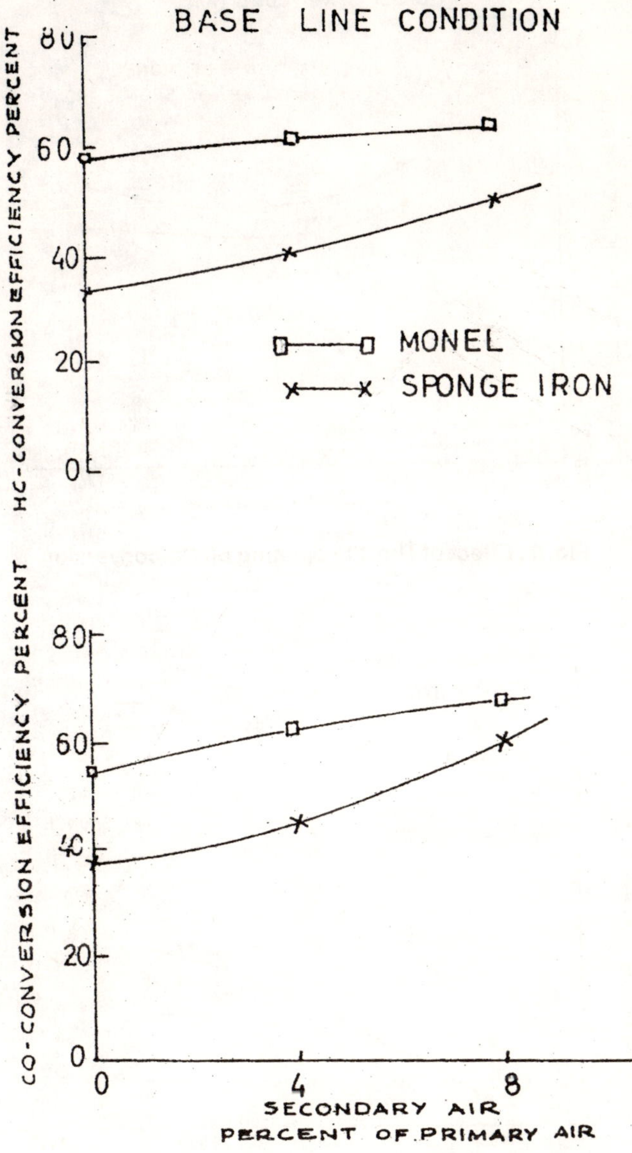

Fig. 7 : Effect of Seconday Air

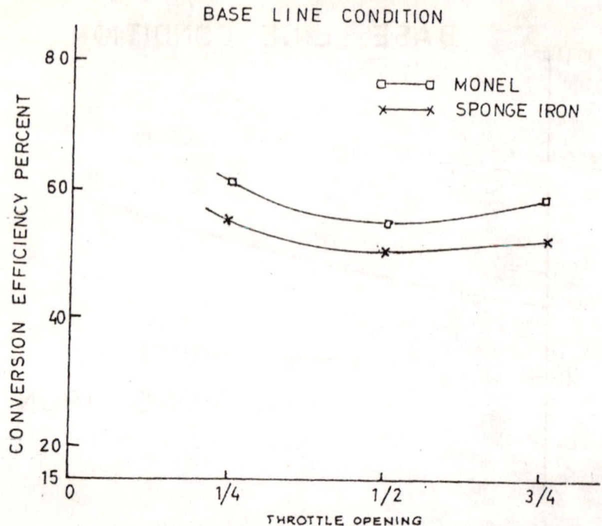

Fig. 8 : Effect of Throttle opening on Co-conversion

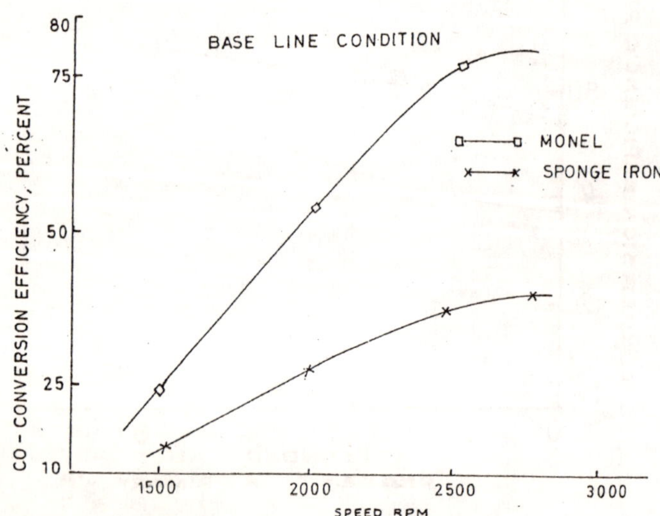

Fig. 9 : Effect of Speed on Co-conversion

Table 2

Speed 2000 r.p.m.Throttle 1/4th full, 10 / Rich Mixture.			
Catalyst	*Conversion*	*Efficiency*	
	CO	*HC*	*NO*
Sponge Iron	65	70	35
Monel	60	55	25

Conclusions

With the large experimental data available both catalysts have characterized the performance regarding reducing the pollutants from S.I. Engine.

The following conclusions could be drawn:

1. The catalytic converter system with sponge iron as catalyst offer less resistance to flow causing less loss of power and gives good response at higher speeds of operation.
2. Sponge iron gave about 65 / reduction in Co compared to 50 / by Monel and for HCs the conversion efficiency was about 70 / with sponge iron compared to 55 / with Monel.
3. As most of the S.I. Engines fitted in automobiles operate-in rich region sponge iron as a catalyst can prove better from the point of performance, availability and durability.
4. Cost of such devices will not come in the way of development of the country, but will definitely add to its economy and healthier conditions.

References

Stern, A.C. (1962). "*Air Pollution*", Vol. I, Academic Press, New York.

Lester, G.R., (1978). Selection of Catalytic Reduction No', SAE paper, 780 202.

Lunt, R.S., Bernstein L.S. et al (1972). Application of Monel Piatinium SAE *Transactions* 720 209.

WU, H.G, Hammertem, R.H. (1983). Development of low cost Thermally stable. Monolithic three way catalyst system. *Ind. Engg. Chem. Prod. Res. Dov.* Vol. 22, No.4.

Rao, S.B. (1986). Catalytic converter system for spark Ignition Engine Exhaust Emission Control-Design Development and Performance Evaluation, Ph.D. Thesis, I.I.T., Delhi.

Sethi, H.M. (1995). Design, Development and Performance of Air Injection System and Catalytic Converter System for small spark ignition engines exhaust emissions control. Ph.D thesis, D.E.I., Agra.

CHAPTER 7

Evaluation of Physico-Chemical Properties of Different Soils Around Cement Producing Areas

Kamal Kishor Kumbhkar and P.S. Dubey
Institute of Environment Management and Plant Sciences
Vikram University Ujjain (M.P.)

Abstract

Three cement producing areas were selected for the impact study of particulate matter input on soil system. The study sites are Vikram Cement (Khor, M.P.), Aditya Cement (Shambhupura, Raj), and J.K. White Cement (Gotan, Raj). These three areas have different soil types i.e. Black cotton, Red laterite and Yellow sandy soil. To evaluate the present status of physico-chemical properties of these, five parameters were selected. A little reduction in the pH of surface soil was observed in comparison to deeper soil layers. In the deeper layer of soil the considerable reduction occur in specific conductivity, salt concentration, osmotic pressure and salinity levels, in comparison to the surface soil. It appears that dust fall settlements affect the surface soil more than the deeper layers of soil.

Introduction

Any entry of foreign particle or polluting agent in the soil system results in alteration to the soil quality. The qualitative changes of soil starts from the physico-chemical changes. And further changes may be biological, physiological, biochemical etc. The physico-chemical characteristics like pH, conductivity, salt concentration, osmotic pressure and salinity are the key factors of the quality of soil.

Particulate matter is a major problem associated with the cement industries which may create a considerable impact on the soil system as well as other related environmental components. Mining and other steps of cement production are also responsible for the enhancement of particulate matter in the adjoining areas of cement industries. Chemically cement dust or its particulate matter contain oxides of Ca, Al, Si, and Na. A little amount of some toxic metals like Cd, Ni, Co, Pb, and Cu are also present in the particulate matter. The present study indicate the degree of impact of cement dust input with reference to the physico-chemical characteristics of different soil types. Three soil types were analysed for this study, which are Red Laterite, Black Cotton and Sandy soil. All the three types of soil exhibit a different response with the addition of particulate matter.

With reference to the cement production the industry consumes 1.6 to 1.8 tonne of raw material for producing one tonne of cement. Obviously there appears to be a significant loss to our mineral resources. In the recent few years a lot of research work related to the environmental consequences of cement industries of the area have been done, (Reddy 1990, Ampily 1992, Sharma 1997, Mankotia 1998 and Dubey 1999). The present study is designed to explore these aspects of cement production at different locations.

Materials and Methods

Three different industrial sites have been selected for this research work in two states which are located in South-West M.P. and border of M.P. and Rajasthan. These three cement industries are Vikram Cement at Khor, (M.P.), Aditya Cement at Shambhupura, (Raj.) and J.K. White Cement Works, at Gotan (Raj.). Various sampling points were selected to soil collection on the basis of wind direction and deflection and meteorological conditions of the study areas. Reference samples were taken at about 15 k.m. away from the industry with no pollution load in comparison to the other sampling sites. Composite soil samples of surface and deeper layers were collected from each sampling point. Surface soil was collected from 0-5 cm. and deeper soil from 25-30 cm depth.

Five parameters were selected for the study of Physico -Chemical properties of Soil i.e. pH, conductivity, salt concentration, osmotic pressure and salinity percentage. For all these parameter pH meters (*Systronics μ pH system 361*) and conductivity meters (*Systronics Conductivity Meter 304*) were used to analyse the samples. The methodology of these parameters is adopted from the *U.S. Salinity Laboratory Staff, 1954.*

Table 1: Physico-Chemical properties of soils at Aditya Cement Area

Sl.No.	*Name of Site*	*pH*	*Conductivity*	*Salt concen- tration*	*Osmotic Pressure*	*Salinity*
			(mMho)	*(me dm^{-3})*	*(bars)*	*%*
1	Reference – S	8.33	0.46	4.6	0.156	0.294
2	Reference – D	8.46	0.38	3.8	0.136	0.243
3	Sawa – S	8.22	0.63	6.3	0.226	0.403
4	Sawa – D	8.32	0.39	3.9	0.140	0.249
5	Kesarpura – S	8.29	0.47	4.7	0.169	0.300
6	Kesarpura – D	8.42	0.41	4.1	0.147	0.262
7	Patniya – S	8.37	0.49	4.9	0.176	0.313
8	Patniya – D	8.46	0.34	3.4	0.122	0.217
9	Kheda – S	8.04	0.49	4.9	0.176	0.313
10	Kheda – D	8.31	0.42	4.2	0.151	0.268

Table 2: Physico-Chemical properties of soils at Vikram Cement Area

Sl.No.	*Name of Site*	*pH*	*Conductivity*	*Salt concen- tration*	*Osmotic Pressure*	*Salinity*
			(mMho)	*(me dm^{-3})*	*(bars)*	*%*
1	Reference – S	8.40	0.40	4.00	0.144	0.256
2	Reference – D	8.44	0.33	3.3	0.118	0.211
3	Khor – S	8.39	0.38	3.8	0.138	0.243
4	Khor – D	8.44	0.35	3.5	0.126	0.224
5	Nayagaon – S	8.38	0.36	3.6	0.129	0.230
6	Nayagaon – D	8.50	0.24	2.40	0.086	0.153
7	Jawad Road – S	8.41	0.43	4.3	0.154	0.275
8	Jawad Road – D	8.46	0.40	4.00	0.144	0.256
9	Kesarpura – S	8.49	0.49	4.9	0176	0.313
10	Kesarpura – D	8.40	0.37	3.7	0.133	0.236

Table 3: Physico-Chemical properties of soils at J.K. White Cement Area

Sl.No.	Name of Site	pH	Conductivity (mMho)	Salt concen-tration (me dm^{-3})	Osmotic Pressure (bars)	Salinity %
1	Reference – S	7.85	0.28	2.8	0.100	0.179
2	Reference – D	7.11	0.63	6.3	0.226	0.403
3	Shivagaon – S	8.11	0.19	1.9	0.068	0.121
4	Shivagaon – D	7.38	0.28	2.8	0.090	0.160
5	Talanpur – S	8.04	0.18	1.8	0.064	0.115
6	Talanpur – D	7.74	0.24	2.4	0.086	0.153
7	Toonklia – S	8.22	0.17	1.7	0.061	0.108
8	Toonklia – D	7.99	0.26	2.6	0.093	0.116
9	Gotan – S	8.36	0.13	1.3	0.046	0.083
10	Gotan – D	8.08	0.16	1.6	0.057	0.10

Table 4: Comparative Physico-Chemical Properties of Soils at Three Different Cement Zone

Sl.No.	Parameter		Sawa-Shambhupura Area		Khor-Nayagaon Area		Gotan Area	
			Not Affected	Affected	Not Affected	Affected	Not Affected	Affected
1.	pH	S	8.33	8.29	8.40	8.49	7.85	8.36
		D	8.46	8.42	8.44	8.40	7.11	8.08
2.	Conductivity	S	0.46	0.47	0.40	0.49	0.28	0.13
	(mMho)	D	0.38	0.41	0.33	0.37	0.63	0.16
3.	Salt Conc	S	4.6	4.7	4.00	4.9	2.8	1.3
	me dm-3	D	3.8	4.1	3.3	3.7	6.3	1.6
4.	Osmotic	S	0.165	0.169	0.144	0.176	0.100	0.046
	Pressure (bars)	D	0.136	0.147	0.118	0.133	0.226	0.057
5.	Salinity	S	0.294	0.300	0.256	0.313	0.179	0.083
	(%)	D	0.243	0.262	0.211	0.236	0.403	0.102

Results and Discussion

Separate reference sites were selected for the sampling of soil in each cement zone because each cement producing area has its own soil type i.e. Red Laterite Soil around Sawa Shambhupura area, Black Cotton Soil around Khor Nayagaon area and Sandy soil around Gotan area. The soil of Gotan area having a higher percentage of sandy materials.

Variations in pH, conductivity, salt concentration, osmotic pressure, and salinity percentage was observed on the basis of soil types and a difference was also noted between the surface and deeper layers of the soil. Among the three soils the pH of Black Cotton soil and Red Laterite soil was higher in the deeper soils while conductivity, salt concentration, osmotic pressure and salinity percentage was low in the sub soils. In case of sandy soil, the pH of deeper soil sample was lower in comparison to the surface soil, but the conductivity and other physico-chemical properties were higher in the lower strata of the soil.

The saturated extract of the fully saline soil has conductivity > 4 mMho, salt concentration > 40.00 me dm^{-3} and pH < 8.00. According to the above definition the conductivity of all three sites soil is < 4 mMho and are apparently free from the problem of salinity. The highest levels of conductivity noted was 0.63 mMho and salt concentration of 63 me dm^{-3} was noted in the sandy soil of Gotan area at reference site.

The osmotic pressure is a pressure exerted in the living bodies as a result of unequal salt concentration on the both sides of the cell wall or cell membrane. Due to this unequal salt concentration, water moves from lower salt concentration to the higher side and therefore, exerts additional pressure on the side with higher salt concentration. The highest value of osmotic pressure was observed in the deeper layers of soil of reference site at Gotan area (0.226 bars) followed by the soil of Khor Nayagaon area (0.176 bars) and soil of Sawa Shambhupura area (0.169 bars).

Table 4 indicate that the conductivity, salt concentration, osmotic pressure and salinity percentage increases in the soils of the nearest sampling points of the industries. All the three soil types follow the above pattern probably due to the increased particulate matter input in these areas.

Acknowledgement

The authors wishes to thanks these industries for the financial support.

References

Ampily, R. (1992). *Study on Certain Air Pollutants and Plant Pathogens interaction.* Ph. D. Thesis. Vikram University Ujjain. M.P.

Dubey, S. (1999). *Heavy Metal and Pesticides Dynamics in Ecosystem Components around Cement Zone.* Ph. D. Thesis, Vikram University Ujjain. M.P.

Fitzpatric, E.A. (1986). *An Introduction to Soil Science. Vol.II* Longman Group Limited, Hong Kong.

Mankotia, A.S. (1998). *Heavy Metal Contamination in Some Ecosystem Components around Industrial Areas.* Ph. D. Thesis. Vikram University Ujjain. M.P.

Reddy, K.V. (1996). *Effect of Cement Dust Pollution on Plants in Neemuch Nimbahera Cement Zone.* Ph. D. Thesis, Vikram University Ujjain. M.P.

Singh, S.N. & Rao, D.N. (1978). Effect of Cement Dust Pollution on Soil Properties and Wheat Plants. *Indian Jr. of Environ. Health.* 20(3): 258-267p.

Sharma, A. (1997). *Studies of Cement Dust and Heavy Metal Impact on Plant System around Cement Industries.* Ph. D. Thesis, Vikram University Ujjain. M.P.

Tyler, G. (1981). *Soil Biochemistry.* Eds. E.A. Paul J.N. Ladd. New York. 5: 371p.

U.S. Salinity Laboratory Staff, (1954). Quality of Irrigation Water. In: *Chemistry of irrigated soils* (eds.)

CHAPTER 8

Effect of Particulate (Cement Kiln Dust) Pollution on *Hibiscus cannabinus* L.

Ch. Uma, T.V. Ramana Rao and J.A. Inamdar

Department of Biosciences
Sardar Patel University
Vallabh – Vidyanagar

Abstract

Variations in the morphology, biochemistry and epidermal features of plants grown under simulated cement kiln, dust polluted environment and control plants are discussed. Pernicious effects on plant growth, phytomass and net primary productivity associated with reduction in chlorophyll 'a', chlorophyll 'b', total chlorophyll, lipids, amino acids, sugars and starch content were observed in dusted plants. However, there is an increase in total phenol content in dusted plants. Epidermal features of the dusted plants showed low frequency of stomata, higher stomatal index and higher density of trichomes. All these morpho-biochemical variations lead to reduced yield in dusted plants.

Introduction

Cement dust, the major source of particulate pollution from cement factories, causes a decrease in crop production by covering the leaves with an incrustation of the dust formed after mixing with mist or light rain (Peirce, 1910). The effect of cement kiln dust on the growth and reduction in phytomass, net primary productivity (NPP), yield are reported by many workers (Uma *et. al.*, 1993; Prasad *et. al.*, 1991; Prasad & Inamdar, 1990; Indhirabai *et. al.*, 1988). The present investigation has been carried out to assess the effect of cement kiln dust pollution on *Hibiscus cannabinus*, an economically important fibre yielding plant.

Materials and Methods

Seeds of *H. cannabinus* were obtained from local market. Seedlings were raised in University Botanical Garden in a plot of 8 sq.m.[2] The seeds were sown at 15 cm intervals in rows with a distance of 10 cms between two rows. The plants in one plot considered as control, whereas those in the other were uniformly dusted with cement kiln dust and treated as dusted plants. Growth analysis and biochemical estimations were carried out in control as well as dusted plants at regular interval of 20 days. Fresh leaf materials were used to determine the following aspects. The method adopted for determining each aspect is given in parenthesis :

Chlorophyll	-	(Mac Lachlan and Zalik, 1963)
Proteins	-	(Layne, 1957)
Starch	-	(Mc Creddy *et al.*, 1950)
Sugars	-	(Dubois *et al.*, 1956)
Lipids	-	(Bolch *et al.*, 1957)
Aminoacids	-	(Moore and Stein, 1948)
Phenols	-	(Bray and Thorpe, 1954)
Phytomass	-	(Kvet *et al.*, 1971)
Net primary productivity		

Epidermal peels have been obtained from the leaflets and stained with ferric alum haematoxylin in for 10 min. each. Stomatal index was calculated following the method of Salisbury (1927, 1932).

Results and Discussion

The dust that accumulates on leaf surfaces adheres to the leaf surface as a coating due to humidity in the dusted plants. The thickness of coating increases with the age of the plants and formation of dust on dusted plant surfaces. The dusted plants showed reduction in their growth as well as the number of branches, number of flowers, number of fruits and also reduction in the length of their shoots, phytomass and NPP. The root/shoot ratio, however, is found to be increased (Table 1). Similar observations were made by Jones and Cowling (1978). Indhirabai *et. al.*, (1988) opined that the reduction in growth might be due to the reduced intensity of light energy available through the cement dust coating for photosynthesis in the dusted plants. A decrease in the phytomass and net primary productivity values of dusted plants is found in direct proportion to the reduced photosynthetic activity. A similar phenomenon has been reported by Darely (1966) and Lerman (1972), who attributed the reduction in the phytomass and NPP is due to the stomatal clogging which results into a reduced gaseous exchange. Thus, it may ultimately lead to the reduction in phytomass and NPP. Klincsck (1975) is of the opinion that cement kiln dust polluted micro environment provides unfavourable ecological conditions for pollen germination and fertilization. The opinion of Klincsck (1975) is found to be true as the control plants of *H. cannabinus* at their 120 days age bore 56.438 fruits, while the dusted plants of same age had only 37.127 with the reduction of 34.216% fruits. Klincsck also observed that pollen requires acidic pH for germination and fertilization. From the available literature, it is found that the plants respond by a lower yielding to a higher rate of pollution (see Uma *et al.*, 1993; Prasad *et al.*, 1991).

One of the characteristic biochemical feature of all the worked out plants that are subjected to cement kiln dust is the reduction in the total chlorophyll. Moreover, reduction in the total chlorophyll content of dusted plants has been reported by several workers (see Bredemann, 1932; Czaja, 1962; Darely, 1966; Lerman, 1972; Singh and Rao, 1978; Pawar *et. al.*, 1982; Prasad *et. al.*, 1991; Uma *et. al.*, 1993). In the present investigation, dusted plants showed reduction in chlrophyll, chlorophyll 'a' and chlorophyll 'b'. This is due to the shading effect of dusted layer (Peirce, 1910; Czaja, 1962; Bohne, 1963).

A decline in the quantity of metabolites such as proteins, amino acids, lipids, sugar and starch content is observed in dusted plants (Table 3). This is probably due to decrease in chlorophyll content. The present observation on the reduction in protein content in dusted plants is in accordance with the observations of earlier workers (Prasad, 1980; Pawar *et. al.*, 1982; Prasad and Inamdar, 1990). In dusted *H. cannabinus* plants the starch content is found to be less in comparison to that of control plants. Bohne (1963) states that the development of cement incrustations may interfere with the light

Table 1: Morphological Parameters of Control and Dusted Plants of *H. cannabinus**

Age of Plant (Days)		Shoot Length (cms)	Root Length (cms)	No. of Leaves	Size of Leaves (cms)	No. of Branches	No. of Internodes	No. of Flower	No. of Fruits	Root/ Shoot Ratio	Phytomass (gm)	NPP (gm)
20	C	10.6± 1.028	8.08± 0.523	8.51± 224	33.10± 1.34	–	4.12± 1.24	–	–	0.762± 0.010	1.253± 0.421	0.062± 0.0014
	D	14.6± 1.104	11.3± 0.576	6.8± 0.126	2.6± 1.304	–	3.24± 1.184	–	–	0.773± 0.002	0.926± 0.243	0.046± 0.0012
40	C	4.61± 0.578	28.3± 0.256	24.8± 1.208	10.21± 2.120	3.4± 1.24	14.401± 1.120	8.124 0.024	–	0.696± 0.012	5.025± 1.024	0.1256± 0.128
	D	35.3± 1.013	26.4± 0.120	20.3± 2.103	8.6± 1.341	2.81± 1.04	12.618± 2.304	3.910± 0.120	–	0.747± 0.024	4.136± 0.248	0.1034± 0.020
60	C	102.8± 0.569	64.6± 0.204	80.6± 2.694	12.4± 0.567	5.34± 1.034	18.420± 1.340	33.120± 1.248	12.124± 1.289	0.632± 0.014	12.243± 2.188	0.2041± 0.106
	D	75.6± 1.029	51.81± 1.381	45.6± 1.384	10.8± 1.246	4.32± 0.248	15.520± 1.268	20.238± 1.234	8.128± 1.128	0.685± 0.13	7.513± 1.218	0.1252± 0.12
80	C	125.5± 2.56	68.4± 1.302	104.8± 1.264	15.5± 1.318	6.47± 1.120	20.413± 1.038	49.278± 1.210	24.121± 2.184	0.545± 0.028	20.146± 1.283	0.2518± 0.0132
	D	102.4± 1.60	64.8± 1.24	57.6± 1.208	12.4± 2.148	5.64± 1.348	18.813± 1.0120	28.138± 1.048	16.216± 1.814	0.632± 0.004	16.823± 2.134	0.2102± 0.024
100	C	147.6±	86.2± 1.318	120.4± 2.318	15.8± 1.012	6.64± 1.0308	24.618± 1.548	54.126± 2.238	38.168± 1.84	0.584± 0.012	31.564± 2.418	0.3156± 0.020
	D	105.3± 1.243	71.4± 1.406	68.9± 1.257	12.8± 1.231	5.84± 1.286	19.84± 1.281	32.235± 1.248	22.146± 2.268	0.659± 0.014	20.213± 1.126	0.2021± 0.024
120	C	181.3± 1.028	94.6± 1.120	127.8± 3.164	15.6± 0.238	6.34± 1.012	26.403± 1.120	41.237± 2.238	56.439± 2.473	0.521± 0.012	39.782± 1.452	0.3315± 0.018
	D	1.304± 1.210	75.4± 1.148	84.8± 1.368	14.9± 1.034	5.44± 1.120	21.416± 1.318	48.128± 1.273	37.127± 2.184	0.578± 0.018	28.125± 1.136	0.2343± 0.034

* Average of 5 replicates. C = Control, D = Dusted, NPP = Net primary productivity.

Table 2: Epidermal Features of Control and Dusted Plants *H. cannabinus**

Age of Plant (Days)		*No. of Stomata*	*No. of Epidermal Cells*	*SI*	*Size of Stomata*	*Length of Trichome*	*Size of Epidermal Cells*	*No. of Trichome*
20	C	154.8± 2.032	438.5± 2.103	26.51± 1.002	0.184± 0.0048	0.50± 0.16	0.448± 0.014	7.23± 0.148
	D	102.2± 1.248	326.4± 1.218	29.74± 1.21	0.016± 0.002	0.305± 0.48	0.040± 0.010	6.82± 1.024
40	C	278.4± 2.013	489.2± 1.224	36.26± 1.043	0.0210± 0.0023	0.463± 0.173	0.0478± 0.012	7.64± 1.203
	D	235.8± 1.307	472.8± 1.015	33.27± 2.180	0.0219± 0.023	0.440± 0.137	0.0443± 0.14	6.93± 1.218
60	C	325.9± 2.104	625.8± 1.2204	34.24± 1.102	0.0281± 0.0034	0.068± 0.010	0.0768± 0.010	7.86± 1.314
	D	298.5± 1.286	528.6± 1.010	36.08± 1.024	0.0327± 0.0030	0.0726± 0.0105	0.0713± 0.012	7.102± 1.146
80	C	332.8± 5.128	640.4± 1.208	34.192± 2.416	0.0317± 0.0030	0.7875± 0.018	0.805± 0.0048	8.15± 1.189
	D	298.6± 2.104	562.3± 1.014	34.66± 1.846	0.0308± 0.0025	0.755± 0.0064	0.0767± 0.0021	7.68± 0.0026
100	C	493.6± 6.102	698.3± 1.219	41.412± 3.120	0.0405± 0.0030	0.792± 0.002	0.865± 0.10	8.24± 1.104
	D	316.4± 1.268	592.4± 1.128	34.815± 1.276	0.0308± 0.0025	0.768± 0.010	0.0767± 0.021	7.83± 0.0026
120	C	525.8± 2.138	728.5± 1.964	41.919± 1.024	0.0425± 0.018	0.0814± 0.012	0.0893± 0.024	8.62± 1.012
	D	415.0± 1.203	658.42± 1.102	38.69± 1.120	0.0378± 0.10	0.0802± 0.014	0.0785± 0.0015	7.92± 1.024

* Average of 20 replicates. C = Control, D = Dusted, SI = Stomatal index.

Table 3: Biochemical Esimtations of *H. cannibinus* Control and Dusted Plants (in mg/gm)*

Age of Plant (Days)		Chlorophyll 'a'	Chlorophyll 'b'	Total Chlorophyll	Total Sugar	Protein	Starch	Lipids	Amino Acids	Total Phenols
20	C	0.874± 0.002	0.463± 0.08	1.337± 0.010	8.426± 1.284	9.710± 0.0016	7.248± 0.0016	8.416± 1.204	7.258± 1.136	3.452± 0.210
	D	0.814± 0.113	0.409± 0.010	1.223± 0.021	7.128± 1.41	8.10± 0.107	6.056± 0.0044	6.942± 1.037	6.342± 1.268	3.0138± 0.424
40	C	1.042± 0.124	0.897± 0.012	1.936± 0.01	10.429± 2.143	11.248± 1.246	9.217 1.124	11.624± 2.184	10.842± 2.62	4.978± 0.124
	D	1.012± 0.120	1.318± 0.106	2.616± 0.024	12.173± 1.248	15.428± 1.340	10.133± 1.812	10.426± 1.624	12.947± 1.403	6.127± 0.624
60	C	1.634± 0.106	1.273± 0.120	2.907± 0.016	18.672± 2.106	17.749± 1.248	12.236± 2.204	14.214± 3.273	14.276± 2.604	5.914± 0.268
	D	1.498± 0.120	1.318± 0.106	2.616± 0.024	12.173± 1.248	15.428± 1.340	10.133± 1.812	10.426± 1.624	10.947± 1.403	6.127± 0.624
80	C	1.994± 0.146	1.284± 0.407	3.278± 0.40	22.764± 1.489	18.263± 2.426	18.721± 1.296	19.427± 2.317	16.923± 2.240	6.863± 1.219
	D	1.615± 0.310	1.215± 0.248	2.888± 0.128	20.814± 2.167	16.829± 1.284	16.124± 2.137	14.173± 1.475	14.297± 2.128	7.120± 0.534
100	C	1.875± 0.210	1.190± 0.104	3.065± 2.210	26.813± 2.184	22.248± 1.236	22.732± 2.178	18.263± 1.7325	18.246± 1.378	5.973± 2.118
	D	1.579± 0.120	1.172± 0.104	2.751± 0.214	24.774± 1.190	20.126± 1.248	20.137± 1.483	15.317± 2.748	16.627± 1.246	6.734± 3.268
120	C	1.684± 0.148	1.010± 0.170	2.703± 0.120	28.784± 1.347	20.426± 2.248	24.848± 3.142	16.326± 2.127	19.487± 2.473	4.918± 0.284
	D	1.463± 0.170	1.126± 0.120	2.589± 0.210	26.124± 2.475	21.126± 1.462	23.126± 2.918	14.128± 1.214	17.261± 1.248	5.164± 0.475

* Average of five replicates. C = Control, D = Dusted.

required for photosynthesis and this reduces the starch production. Many other workers (e.g. Darley, 1966; Lerman, 1972) have also come to the conclusion that the critical factor in starch production is a high degree absorption of heat in cement kiln dusted leaves due to limited transpiration. This hinders the phosphorylation of sugars and translocation from assimilating leaves.

Furthermore, the quantity of total sugars, lipids and amino acids is also reduced in the dusted plants of *H. cannabinus* comparison to that of control plants (Table 3). Ting and Mukerji, (1971) reported a reduction in total sugars content in dusted plants and opined that it may be due to reduced photosynthetic rate. However, increasing of total phenol content was a remarkable defence mechanism.

The epidermal features of dusted leaves exhibit decrease in no. of their epidermal cells and stomatal frequency, but there is an increase in stomatal index and trichome frequency (Table 2). Similar observations have been made by Prasad and Inamdar (1990) also. A decrease in the number of stomata in leaf epidermis of dusted plants indicates a favourable adaptation. Sharma and Butler (1973) and Yunus and Ahmed (1980) opined that this kind of adaptation may regulate the transpiration as well as to limited and controlled the entry of harmful pollutants into plant tissues. Moreover, increase in the number of trichomes would protect the leaf from direct exposure to sun rays and pollutants.

All these morpho-biochemical and epidermal variations indicate that there an impact of cement kiln dust pollution on the growth of *H. cannabinus*, which ultimately lower downs its yield.

References

Bohne, H. (1963). Schadlichkeit von stub ans zementwerken fur wald be stande. *Allq. Forstz.* 18: 107-111p.

Bray, H.G. and Thorpe, W.V. (1954). Analysis of phenolic compounds of interest in metabolism. *Meth. Biochem. Anal.* 1: 27-57p.

Bredemann, G. (1932). Botanische untersuchungen bei Rauschaden. In: E. Haslhoff. G. Bredemann und W. Hazelhoff, 1932, "Entstenung, Erkennung und Beurteilung von Rauschaden". Gebr. Borntraeger verl., Berlin, 285-392p.

Czaja, A.T. (1962). Uber das problem der zements stubwirkungen auf Pflanzen. *Staub.* 22: 228-232p.

Darely, E.F. (1966). Studies on the effect of cement kiln dust on vegetation. *J. Air Pollu. Contr. Ass.* 16: 145-150p.

Dubois, M., Gilles, K.A., Hamilton, J.K., Roberts, P.A. and Smith, F. (1956). A colorimetric method for determination of sugars and related substances. *Analyt. Chem.* 28: 350-356p.

Folch, J., Lees, M. and Sloan Stanley, G.H. (1957). A simple method for the isolation and purification of total lipids from animal tissues. *J. Biol. Chem.,* 226: 497-509p.

Indhirabai, K., Dhanalakshmi, S. and Lakshmanan, K.K. (1988). A significant observation on the effect of cement, cement dust, H_2S and NaCl on growth and development of Dolichos biflorus L., *Adv. In Pl. Sci.,* 11: 91-101p.

Jones, L.H.P. and Cowling, D.W. (1978). Effects of air pollution, Part 2 – Plants and farm animals – General principles, specific poilutants, rural industries and air quality. In "*Industrial Air Pollution*". *Hand Book* (ed.) Parker, A., McGraw-Hill Book Company, London. 32-56p.

Kvet, J., Ondok, J.P., Necas, J. and Jarvis, P.G. (1971). *Manual methods of plant photosynthetic production.* Dr. W. Junk, N.V. Publishers. The Hague.

Klincsek, P. (1975). *Response of Ornamental trees and shrubs to pollution by chemically active dusts.* Doctoral dissertation, MTA Budapest.

Layne, E. (1957). Spectrophotometric and turibdimetric methods for measuring proteins. In "*Methods in Enzymology*" (ed.) Colowack, S.P. and N.O. Kaplen, Academic Press, New York. 3: 447-466p.

Lerman, S.L. (1972). Cement kiln dust and the bean plant (*Phaseolus vulgaris*) L. Black valentine var.; *Indepth investigations into plant morphology, physiology and pathology.* Ph.D. dissertation, University of California, Riverside.

Mac Lachlan, S. and Zalik, S. (1963). Plastid structure, Chlorophyll concentration and free amino acid composition of a chlorophyll mutant of barely. *Can. J. Bot.* 41: 1053-1062p.

McCreddy, R.M., Guggler, J., Suviera, V. and Owens, H.S. (1950). Determination of starch and amylose in vegetables. *Anal. Chem.*, 22: 1156-1158p.

Moore, S. and Stein, W.H. (1948). Photometric method for use in the chromatography of aminoacids. *J. Biol. Chem.* 176: 367-388p.

Pawar, K., Trivedi, L. and Dubey, P.S. (1982). Comparative effects of cement coal dust and fly ash on *H. abelmaschus. Intern. J. Envi. Studies.* 19: 221-223p.

Peirce, G.J. (1910). On effect of cement dust on orange trees. *Plant World.* 13: 288-291p.

Prasad, B.J. (1980). *Phytotoxicity of refinery air pollutants.* Ph.D. Thesis, Banaras Hindu University, Varanasi.

Prasad, M.S.V. and Inamdar, J.A. (1990). Effect of cement kiln dust pollution on black gram [Vigna mungo (L.) Hepper] *Proc. Indian Acad. Sci.* 100: 435-443p.

Prasad, M.S.V., Subramanian, R.B. and Inamdar, J.A. (1991). Effect of cement kiln dust on *Cajanus cajan* (L.) Millsp. *Ind. J. Env. Hlth.* 33: 11-21p.

Salisbury, E.J. (1927). On the causes and ecological significance of stomatal frequency with special reference to the wood land flora. *Phil. Trans. R. Soc. B.* 216: 1-65p.

Salisbury, E.J. (1932). The interrelation of soil, climate and organism and the use of stomatal frequency as an integrating index of the water relation of the plant. *Beih. Bot. Zentralb.* 49: 408-420p.

Sharma, G.K. and Butler, J. (1973). Leaf cuticular variatios in *Trifolium repens* L. as indicators of environmental pollution. *Environ. Pollut.* 5: 287-293p.

Singh, S.N. and Rao, D.N. (1978). Effect of cement dust pollution on soil properties and on wheat plants. *Ind. J. Environ. Hlth.,* 20: 258-267p.

Ting, J.P. and Mukerji, S.K. (1971). Leaf ontogeny as a factor in susceptibility to O_3 aminoacids carbohydrate changes during expansion. *Am. J. Bot.* 58: 497-504p.

Uma, Ch., Rao, I.V.R. and Inamdar, J.A. (1993). Morpho-biochemical changes due to the cement kiln dust pollution in mustard (*Brassica juncea* Linn.) *Indian Botanical Contactor,* 10: 131-135p.

Yunus, M. and Ahmad, K.J. (1980). Effect of air pollution on leaf epidermis of Psidium guajava L. *Ind. J. Air Pollut. Contr.* 3: 62-67p.

CHAPTER 9

Study of Environmental Impact Variables for Highway Projects

R.D. Bhatt and B.D. Vashi

Civil Engineering Department
Faculty of Technology & Engineering
M S. University of Baroda, Vadodara

Introduction

Roads have been considered as essential infrastructure for Socio Economic growth of the community. On one side, it helps as growth variable by providing quick inputs and outputs of industrial growth, agricultural development, social services, national unity, employment generation, on the other hand, it has impacts in the form of soil erosion, restriction in natural drainage, increase in traffic noise, dust, air pollution, disturbance of Ecology etc. The factors initiating these impact can be due to highway (Visual intrusion, severance, land consumption and change in access etc.), traffic factors (noise/air pollution and vibration) noise and air pollution (exhaust and gases from construction equipment and dust resulting from construction operations).

Impact Assessment

The methodology for identifying and deciding weightage to each impact has been developed at various stages in a process of using EIA has an integral tool of assessment of highway project.

The Environment Impact Assessment (EIA) involves (i) prediction of environmental impact of project (ii) determining ways and means to reduce adverse impacts (iii) Shaping projects to suit the local environment and (iv) Present the predictions and options to the decision makers. Prediction of these impacts have been difficult process as many of these impacts are:

1. *Positive impacts* helping the community to grow as well as *Negative impacts* in terms of soil and drainage erosion, disturbance to ecology and wild life, changes in the value system as well as initiating increase in air, noise and dust pollution.
2. It involves tangible (which can be converted in terms of money) as well as intangible impacts and
3. Direct to the users to indirect (even to non-users of the facility)

Planners decision involves taking all the impacts to a common parameter for comparison. For this many methodology and steps have been evolved in a process of framing EIA statements over a period of time. Many of these parameters can be classified to subjective as well as objective category these when applied for Indian conditions can be used for directing with little or no modification of parameters, while many of them still needs improvement and calibration for further use.

Various Methodology for EIA of Highway Project

In this paper, methodology for some of the EIA statement as reported in the literature and as used for professional or research work for applying to Indian conditions is being discussed.

Application of Leopold statement for overall Impact Assessment

Leopold stated that an assessment of the probable impacts of the variety of the specific aspects of the proposed development and an estimate of the magnitude of each:

(i) a listing of the effects on the environment which would be caused by the proposed development and an estimate of the magnitude of each
(ii) an evaluation of the importance of each of these effects
(iii) the combining of magnitude and importance estimates in terms of a summary evaluation.

For each scheme being assessed these three elements are analysed using a matrix, on one axis of which are the actions which are existing environmental actions which may cause environmental impact, grouped into modifications of the regime, land transformation and construction, resource extraction, processing, land alteration, resource renewal, changes in traffic, waste emplacement and treatment, chemical treatment, and accidents. Listed vertically are 88 environmental characteristic grouped into physical and chemical characteristics, biological conditions, cultural factors, and ecological relationships.

In a research project of expressway link between Baroda, and Ahmedabad by L.D. College of Engineering. Ahmedabad, this technique has been used. Also they have used matrix developed by Lohani and Thann based on assigned values as adopted by the following equation –

Σ priority value x Magnitude x Significance

Environmental Quality Designation (EQD)

Delphi technique for ecology, environment pollution and human interest for a hypothetical road construction project can be explained as under –

Let us say the rating for Ecology be X, Environment Pollution be Y and Human interests be Z.

Then X+ Y + Z = 100

Let us say that the sum of rating given to Ecology be $\sum_{i=1}^{n} Xi$, Environmental Pollution be $\sum_{i=1}^{n} Yi$ and Human interests be $\sum_{i=1}^{n} Zi$, where n = No. of elements considered in these three sub-system.

Then Environmental Quality Designation

$$EQD = \frac{\sum_{i=1}^{n} Xi}{X} \times \frac{\sum_{i=1}^{n} Yi}{Y} \times \frac{\sum_{i=1}^{n} Zi}{Z}$$

The values of X, Y and Z can be identified from Table 1 given below for a particular Highway Project. Also values of Σ Xi, Σ Yi and Σ Zi can be chosen.

Table 1

Value of EQD	Environmental Impact Rating
< 20	Very low
20-30	Low
30-40	Low –Medium
40-50	Medium
50-60	Medium High
60-70	High
>70	Very High

Let for a Typical highway development project the importance rating given to ecology be 40, environmental pollution be 25 and human interest be 35, respectively

Then, X = 40, Y = 25 and Z = 35

Also, X + Y + Z = 100

The values of ΣX, ΣY and ΣZ (Chosen from tables) be taken as 70, 45 and 42 during the activity and 85, 50 and 60 after the activity is complete.

The EQD value during the activity

$$= \frac{70}{40} \times \frac{45}{25} \times \frac{42}{35}$$

$$= 3.78$$

Which is less than 20 which indicates low impact.

EQD value after the activity

$$= \frac{85}{40} \times \frac{50}{25} \times \frac{60}{35}$$

$$= 7.27$$

Which is still less then 20 indicating low impact. But it can be seen that after activity EQD has increased much.

Air Pollution Measurement Study

In one of the study conducted by IIT, Bombay for expressway link between Baroda-Ahmedabad Gaussian Model and Air Pollution Index have been employed to product the pollutants of receptor level, 3 m away from the road side kerb.

(a) The Gaussian line source formula :

$$c(X) = \sqrt{2/\Pi} * q / [\sin(\varnothing)\, \partial_{Z(X)}\, \mu]^* \exp[-½ (H / \partial_{Z(X)})]$$

where

$C(x)$ = air pollutant concentration at a receptor located at distance x from the highway in the downwind direction ($g\ m^{-3}$)

q = line source strength ($g\ see^{-1}\ m^{-1}$)

ϕ = angle between wind direction and the line source.

$\partial_{Z(X)}$ = the vertocal standard deviation or plume concentration distribution at the downwind distance x for the prevailing atmosphere stability (m).

μ = mean wind speed ($m\ sec^{-1}$)

H = effective height or emission of the line source (m).

(b) Air Pollution Index

There are a Number of Indices reported in literature.

Oak Ridge Air Quality Index (ORAQI) is being used here. ORAQI index may be calculated for any combination of pollutants from i = 1 to 5 using momogram. Pollutant standard Si are 24 hr. extrapolation of secondary NAAQA (National Ambient Air Quality Standard). When pollutant concentrations are at background levels, ORAQI = 10, when all concentrations are at standards ORAQI = 100.

$$ORAQI = 5.7 \left[\sum_{i=1}^{5} (Ci / Si) \right]^{**} 1.37$$

Where

i = Variables (CO, SO_x, TSP, NO_x, O_x)

Ci = Concentration of pollutant

Si = Standard for pollutants

Si Values for the five pollutants are given below :

(a)	Carbon monomide	7800 $\mu g/m^3$
(b)	Nitrogen Dioxide	400 $\mu g/m^3$
(c)	Photochemical oxidants	59 $\mu g/m^3$
(d)	Particulate matter (TSP)	150 $\mu g/m^3$
(e)	Sulphur dioxide	266 $\mu g/m^3$

At present no indigenous data or models are available for predicting the level of ambient air pollution under given sets of conditions.

Till such time data/models are available it is suggested by I.R.C. that both the number of motorised vehicles and the volume/capacity ratio (lower the ratio, lesser the congestion and lower pollution) be taken as indicators for air pollution levels.

Prediction of Noise Levels

The major factors which influence the generation of road traffic noise are :

(a) the traffic flow

(b) the traffic speed

(c) the proportion of heavy vehicles

(d) the gradient of the road

(e) the nature of the road surface

In addition the following factors influence the noise level at a reception point distant from the highway –

(f) attenuation of the around waves due to distance between source and receiver and also due to ground absorption

(g) construction of the sound waves by buildings or noise barriers

(h) construction of the sound waves due to a restricted angle or view of the source line from the reception point.

(i) reflection effects

By using nomograms developed by Department of Transportation memorandum, calculation of road traffic noise, the basic noise level in terms of the 18 hrs. (6 a.m. to midnight) traffic flow for a normalised source to receiver distance of 10 mtr. When the mean traffic stream speed is 75 km/h. there are no heavy vehicle in the flow and roadway is in level. The traffic flow (Q) is the two way flow from 6 a.m. to midnight.

The relationship between L_{10} (18h) basic noise level and traffic volume expressed mathematically as

$$L_{10}(18h) = 29.1 + 10 \log Q \text{ d B A}$$

Where L_{10} is the sound pressure level exceeded for 10% of the time.

Another set of nomograms have been suggested by The National Co-operative Highway Research Programme (NCHRP). U.S.A. In a Revised Design Guide (RDG), it is suggested,

(i) $L = 18 + 30 \log S$ d B A (for Automobiles),

(ii) $L = 86$ d B A (For trucks)

Where

L	=	Loudness emission level in 'noys'
S	=	Speed of vehicle in MPH
dBA	=	Sound frequency (for annoyance) for the A Curve which is based on 40 phon contour.

With the available data it is not possible to predict noise levels for a given set of road and traffic conditions. However it may be judged subjectively that a road with lower traffic, lower volume/capacity ratio and smoother surfacing will develop lower noise levels than the one otherwise. Where higher noise are experienced or anticipated it would advisable to provide a screen plantations and other noise attenuating devices.

In India some work has been reported in literature for evaluating the economic significance of transport proposal.

In his study, submitted to M.S. University for M.E. Degree, Khalid (1984) has tried to develop geo-climatic region concept in reference to road project.

Conclusion

There is necessity of evaluating efficiency of various methodology in Indian context, which requires further work in this direction.

References

IRC : 104:1988 and FHAJ guidelines for preparation of EIA statements.

Louis F. Cohn and Gary R. Mevoy : Environmental Analysis of Transportation systems. John Willey & Sons Publications.

Environmental impact assessment of Ahmedabad, Baroda express way project report. L.D. College of Engineering, Ahmedabad. May 89.

Indian Highways (IRC Publications) Volume

Volume 18 No.12 Dec. 1990 (Page 5 to 9) V

Volume 19 No.4 April- 91 (Page 5 to 9)

Volume 20 No.4 April -92 (Page 21 to 28)

Volume 20 No.6 June 92 (Page 7 to 10)

Khalid Ahmed "Significance of Geo environmental controls on Road performance: A case study for Baroda - Kadipani Road Link". Dissertation submitted to M.S. University of Baroda in August 1984.

CHAPTER 10

Strategies to Control Emissions in Diesel Driven Vehicles

R. Arun Nirmal, J. Chandrasekar and K. Balagurunathan
Mechanical Engineering Department
College of Engineering, Guindy,
Anna University, Chennai

Abstract

An increase in number of diesel driven vehicle will adversely change the chemical composition of air. The most obvious pollutant from diesel is the visible smoke, which is primarily the carbon portion of the diesel particulate resulting in opacity in urban and rural areas. Drastic reduction of NO_X can be possible at heavy load regions with the combined use of EGR and methanol fumigation.

Introduction

Diesel engines emit different pollutants than do gasolene engines. Major growth in the number and use of diesel vehicles will adversely change the chemical make up of air. The major components of the diesel emissions are particulates, NOX, hydrocarbons, oxides of sulphur, elemental carbon and soluble organic fractions. Particulates are particles of solid or liquid matter. Diesel particulate emissions are less than 2.5 μ. Diesel particulates, consist largely of a solid carbonaceous core upon which is absorbed a complex mixture of oxygenated hydrocarbons (soluble organic fraction), sulphates and trace elements. Heavy duty diesel vehicles emit between 1.5 to 2.0 g/ml particulates.

The most obvious pollutant from the diesel is the visible smoke which is primarily the carbon portion of the diesel particulate which obscures light causing opacity in both urban and rural areas. Most diesel particulates are formed when the fuel is incompletely burned. The diesel particulates consist of clusters of individual spheroids which are fused together to form the primary particle. These primary particles are non-spherical and have a mean diameter of 0.1 to 0.2 μ. This fine sized particle has a long lifetime in the atmosphere and is capable of being inhaled deeply into the lungs causing long term health hazards.

In order to meet the requirements, each country specifies the emission allowed from the tail pipe of different types of vehicles. Vehicles powered by diesel engines must by 1994 and beyond meet very stringent requirements of NO_x, hydrocarbons and particulates. Hydrocarbon is fairly easy to meet; but to meet the NO_x and particulate requirements requires much trial and error engine modifications, such as trying very high injective pressures, and other things. Shock wave generation around the diesel fuel sprays with high pressure injection is one of the latest technologies developed to reduce smoke and NO_x emissions in Dl diesel engines by generating strong turbulence in the later stages of combustion.

Shock wave generation around the diesel fuel spray with high pressure injection

While studying the characteristics of the high pressure fuel spray which is injected in a high pressure vessel, Toshio Nakahira *et al.* found weak shock waves generating around the fuel spray. To investigate the shock wave effect on the fuel spray, Toshio Nakahira *et al* measured the propagation speed and pressure amplitude as function of the injection pressure and ambient pressure. The results indicated that shock waves are generated when the fuel injection speed exceeds the ambient sonic speed. Also it was found out that the pressure amplitude of the shock wave is approximately 10% of the ambient pressure and the shock waves spread at sonic speed. It is therefore possible to use the shock waves for combustion improvement and for reducing emissions.

NO_x Reduction from diesel combustion using pilot injection with high pressure fuel injection

Shigeru Shundoh *et al.* tested several methods to reduce the ignition delay period. Heat insulation, ignition assistance by glow plug, various compression ratios and pilot injection were tested combined with high pressure fuel injection and the effects on combustion and emissions were examined in experimental engines. The results obtained reveal that:

1. The shortening of ignition delay does not affect so much at the usual injection timing before TDC, because original ignition delay is so short that the portion of the initial combustion is very small in the case of high pressure injection with small nozzle hole.
2. When the injection timing is retarded or when the original ignition delay is relatively long, shortening of the ignition delay is effective to reduce initial combustion and NO_x emissions.
3. The considerable retard of injection timing is possible by using pilot injection. In this case different tendency from usual injection timing can be seen, such that higher swirl ratio or increased number of nozzle hole give better performance and emissions.
4. The combination of pilot injection and injection pressure simultaneously reduces NO_x by approximately 30% and smoke by 60-80% without worsening the fuel economy.

Reduction of NO_x, smoke BSFC and maximum combustion pressure by low compression ratios in Diesel engines fuelled by Emulsified fuels

Minoru Tsukahara and Yasufumi Yoshimoto of Joetsu University of Education investigated the performance of a 4 cycle water cooled, single cylinder diesel engine having a bore X stroke 110 x 150 mm, and stroke volume 1425 CC, using water-in-gas oil emulsified fuel and gas oil at compression ratios of 13.6, 15.6 and 17.0.

From the data collected it was found out that there was a remarkable reduction of NO_x concentration and smoke density suppressing the combustion pressure rise without worsening the specific fuel consumption, when the diesel engine operates with emulsified fuel at low compression ratios.

The above experiment compared a normal compression ratio of 17.0 for gas oil operation with a low compression ratio of 13.6 for emulsified fuel. The NO_x concentration was 550 ppm with gas oil versus 350 ppm with emulsified fuel, a reduction of about 36%, the smoke density was 5.6 vs. 3.3, a reduction of 41%, and the maximum combustion pressure decreased by 11% without worsening the specific fuel consumption.

Reduction of smoke and NO_x by strong turbulence generated during the combustion process in DI Diesel Engines

Mitsuru konno *et al* conducted a series of experiments in a 1425 cc, four stroke, single cylinder, naturally aspirated direct injection diesel engine with a compression ratio 17.4. A small auxiliary chamber and fuel injection nozzle were installed at the cylinder head of the basic engine to generate turbulence. The results obtained in the above investigations show that:

1. Strong turbulence generated late in the combustion period effectively reduces smoke and particulate without increasing NO_x. Thermal efficiency is proved at high loads because of a shortening of the combustion period.
2. With the present system of turbulence generation, the auxiliary fuel amount must be controlled to suit the engine load to avoid a deterioration in fuel consumption and hydrocarbon emissions of high loads. Smoke reduction becomes larger with higher gas jet momentum. Smoke reduction correlates with the momentum of the gas jet from the turbulence cell (CCD).
3. A combination of EGR or water injection with the CCD system is very effective to achieve a simultaneous reduction in both smoke and NO_x. With the present CCD system, water should be injected at high loads and EGR applied at low loads.

Optimising control of NO_x and smoke emissions from DI Engine with EGR and Methanol fumigation

Matsuo Odaka *et al* of Ministry of Transport, Japan made an attempt to optimize NO_x and particulate emissions from heavy duty diesel powered vehicles under heavy load engine operating regions by combining EGR and methanol fumigation.

The effects on exhaust emissions were experimentally studied and the following results were obtained:

1. Smoke concentration is decreased and total fuel consumption is improved according to the increase in methanol energy ratio.
2. NO_x reduction effect of EGR was not affected by methanol fumigation.
3. Thereby, drastic reduction of NO_x can be possible at heavy load regions with the combined use of EGR and methanol fumigation.
4. NO_x mass emissions were reduced to almost one half of the load without increase in particulate emissions.

Conclusion

The various strategies discussed above require considerable costly modifications of the present Diesel Engines. It would be rather a Herculean task for the Indian Automobile Industries with the present infra-structure to go into the immediate production of such expensive automobiles and market them at home: but it may be possible to initially switch over to the filter trap method of controlling diesel exhaust particles. In filter trap methods, diesel particulate filters such as the mullite corrugation type and the honey comb type of ceramic filters of various small micro-pore sizes and thin wall thickness are used to make a moderate attempt to control particulate emissions. It requires further more well organized dedicated research in the design and development of diesel driven vehicles with controlled emissions before mass production of such vehicles, in India perhaps by the advent of the 21st century.

References

Colorado Diesel Inspection and Maintenance Programme - Jerry Gallagher. SAE 1992.

NO_x reduction from Diesel Combustion using pilot injection with high pressure fuel injection -Shigeru Shundoh, Masanori Komori, Kinji Tsujimura and Shinji Kobayashi. SAE 1992.

Optimizing control of NO_x and smoke emissions from DI Engine with EGR and Methanol Fumigation -Matsuo Odaka, Noriyuki Koike, Yujirou Tsukamoto, and Kazuyuki Narusawa. SAE 1992.

Reduction of NO_x, smoke, BSFC, and maximum combustion pressure by low compression ratios in a diesel Engine fuelled by emulsified fuel. Minoru Tsukahara and Yasufumi Yoshimoto. SAE 1992.

Reduction of smoke and NO_x by strong turbulence generated during the combustion process in DI Diesel Engines -Mitsuru Konno, Takemi Chikahisa and Tadashi Murayama. SAE 1992.

Shock wave generation around the diesel fuel spray with High pressure injection -Toshio Nakahira, Masonori Komori, Masahiko Nishida and Kinji Tsujimura. SAE 1992.

Trapping performance of Diesel Particulate Filters - Osamu Shinozaki, Eiichi Shinoyama and Keizo Saito. SAE 1990.

Chapter 11

Fungal Biodeteriogens Associated with Mummy in Museum and Picture Gallery, Baroda

Arun Arya[1], A.R. Shah[2] and S. Sadasivan[2]

[1] *Botany Department, Faculty of Science, The M.S. University of Baroda, Vadodara - 390002*

[2] *Museum and Picture Gallery Baroda, Vadodara - 390018*

Abstract

There are a number of biotic and abiotic agencies like pollution, light, humidity, temperature, fungi, bacteria, actinomycetes, insects, algae etc. that have a deteriorating effect on museum materials. Of these agencies biological agents such as micro-organisms and insects may cause devastating damages. Except for physical damage by accident, fire, flood or earthquake no other agency is as quick and has a far-reaching consequence in causing serious damage to museum objects as are the biological agents.

A study of aeromycoflora of Museum and picture gallery, Baroda was undertaken during 1999. Samples were taken from air, dust present in coffin box and body parts of Egyptian mummy (dated about 230 BC). Twenty five viable species of fungi belonging to 15 genera were isolated from indoor air of Egypto-Babylonian gallery by gravity fall method. The isolation resulted in 6 different fungi from mummy chamber and dust belonging to Zygomycotina, Ascomycotina and Fungi imperfecti. The mycoflora was dominated by different Aspergilli. Isolations from the first finger of right leg of 1.54m long Egyptian mummy revealed the presence of *Emericella nivea* Wiley & Simmons. This fungus produced white coloured patches on the fingers of mummy. In culture both asexual and sexual stages were observed. The paper deals with studies on mycodeteriogens present in indoor air and fungi affecting one of the seven unique museum objects present in our country.

Introduction

The Baroda Museum and picture gallery is famous for its collection of European oil paintings all over Asia. The buildings of museum are situated in pastoral environment of Sayaji park. This museum was planned by Maharaja Sayajirao Gaekwad III in Indo-Saracenic style of architecture. It was thrown open to public in 1894. Baroda museum is one of the 7 museums in the country where Egyptian mummies are kept for display. The mummy placed here was originally obtained from Cairo (Egypt). It was supplied to the Baroda museum by H.A. Ward of New York, U.S.A on 21st December 1894.

The biopollutants like fungi, bacteria and actinomycetes pose severe threat to museum objects. Biodeterioration of cultural property may result by the activity of these notorious micro-organisms. Microbes get entry in to indoor atmosphere through wind currents and settle on various objects by impaction. These are disseminated by the action of wind. At the threshold of new millennium when we are concerned to conserve these valuable links of human civilization the detailed studies are required on deteriorating behaviour of these micro-organisms. These rare specimens if well preserved will help to educate the generations yet to come about human life existed in ancient past.

It is interesting to note that any biodeteriorating object may act as a small or large bio-industry, emitting neither polluted water nor noxious gases but large number of bio propagules with a potential to start another such unit in a short span. These serve as a secondary source of biodeteriogens. Study on indoor aeromycoflora has attracted the attention of numerous aerobiologists. Gregory (1973), Knox (1979), Spieksma (1980), Arya *et al.* (1988), Chanda (1992), and Shah *et al.* (1996) have emphasized the importance of monitoring such biopollutants in different indoor environments. Considering the need to monitor such biopollutants in the atmosphere of Baroda Museum and picture gallery an effort has been made by Arya *et al.* (1988), Shah *et al.* (1994), and Shah *et al.* (1996). Present research paper deals with assessment of indoor aeromycoflora and occurrence of fungus on deteriorating mummy in mummy chamber kept in Egypto-babylonian gallery.

Materials and Methods

Air sampling has been successful in revealing the diversity of organisms forming the aerospora. Gregory (1973) suggested different air sampling techniques based on

(i) Gravity sedimentation methods

(ii) Impaction

(iii) Use of impingers

Hirst (1952) developed an automatic volumetric spore trap. Since the qualitative studies on aeromycoflora were conducted the petridish exposure method involving sedimentation principle was preferred. Culture plate exposure method (Hyde and Williams 1961) was selected as gravity slide exposure method is found to have certain drawbacks (Nigam and Pathak, 1991) like:

1. Similar spores are not identified up to generic level (e.g. *Aspergillus, Penicillium, Mucor* and *Rhizopus* etc.).
2. Some spores are too small or too transparent to be differentiated clearly or they do not have sufficient characteristic e.g. yeast and *Phoma* etc.
3. Fungal hyphae are not identified.
4. Many spores are not trapped due to strong wind currents
5. This method gives a total spore count, nonviable spores are also counted.

A mycological analysis of indoor air in the Baroda Museum was done at 2 places on 16th April 1999 and 22nd September 1999. Three petriplates (100 mm) each were exposed in the Egypto-Babilonian gallery. The petridishes were also exposed in the mummy chamber for 10 min. They were incubated at 25 + 1°C for 7 days. As soon as fungus showed growth sub-culturing was done. Pure cultures were maintained on PDA slants. Later on these cultures were stored in refrigerator at 8-10°C. Subculturing was done at the interval of every 60 days. Fungi from dust were isolated by dilution plate method (Timonin, 1941). Species of *Aspergillus* and *Penicillium* were identified by growing them in Czapek-Dox's medium and studying their morphological characters. The fungi were stained by method suggested by Booth (1971), and they were photographed. Occurrence of fungi in Baroda museum is recorded in Table 1.

Table 1: Indoor Aeromycoflora of Egypto-Babylonian Gallery in Baroda Museum and Picture Gallery in September 1999

Sl. No.	*Fungi Isolated*	*Colonies**	*Per cent*
Zygomycotina			
1.	*Mucor hiemalis*	8	4.8
2.	*Rhizopus nigricans*	10	6.1
Ascomycotina			
3.	*Aspergillus flavus*	5	3
4.	*A. japonicus*	7	4.2
5.	*A. niger*	17	10.3
6.	*A. ustus*	5	3
7.	*A. wentii*	7	4.2
8.	*Chaetomium atrobrunneum*	6	3.6
9.	*C. gangligerum*	10	6.1
10.	*C. globosum*	5	3
11.	*Emericella nidulans*	5	3
12.	*E. variecolor*	2	1.2
13.	*Lewia infectoria*	5	3
14.	*Monascus rubur*	5	3
15.	*Penicillium citrinum*	7	4.2
16.	*P. nigricans*	5	3
Fungi imperfecti			
17.	*Alternaria alternata*	4	2.4
18.	*Cladosporium herbarum*	2	1.2
19.	*Curvularia lunata*	10	6
20.	*C. verruculosa*	2	1.2
21.	*Fusarium oxysporum*	8	4.8
22.	*Monilia sitophila*	5	3
23.	*Phoma nebulosa*	8	4.8
24.	*Trichoderma ligonorum*	2	1.2
25.	*T. viride*	5	3
26.	Sterile white mycelium	10	6
27.	Sterile black mycelium	5	3
	Total no. of fungal colonies	170	

* The data represents the total nos. of colonies appeared on 5 PDA plates after 7 days of incubation.

Per cent = Percentage occurrence.

Results and Discussion

Mummification: Egypt's Most Mysterious Art

Egyptian society believed not only in existence of God and living beings but also in dead. Their concept was that the soul *Ba* could enter the body again. It could not survive without the preserved body. The body of dead should not deteriorate or it should be attractive enough to lure back soul. Mummification has been tried in past by different communities of different continents. From Greco-Roman period bitumen began to be used in the process and it was the Persian word *Mumiyah* for the bitumen that gave the term *Mummy*. Herodotus a much traveled historian and geographer has given the following account about mummification.

"The mode of imbalming includes taking out of brain by a crooked piece of iron, cutting the abdomen with an Ethiopian stone, taking out of the whole contents of abdomen, which they then cleanse, washing it thoroughly with palm wine. The body is filled with purest bruised myrrh, with cassia and every sort of spicery except frankincense, and sew up the opening. Then the body is placed in natrum for seventy days and then it is washed and wrapped around from head to foot with bandages of fine linen cloth, smeared over with gum. The beautiful drawings are made to depict the details of dead person," (Thorwald, 1962).

In another method syringes were filled with cedar oil and injected in the body without cutting it. The body is placed on salt or natrum for prescribed days then the oil is drained off, which dissolves all the inner body parts. Thus nothing is left of the dead body but the skin and the bones.

The innumerable mummies preserved for thousands of years in mastabas, pyramids, royal tombs, or simply dry earth prove that Herodotus' account, though it refers to the late period, is to some extent applicable to the whole span of Egyptian history. In spite of all the modifications undergone by the Egyptian religion, one basic notion remained. To put it in summary and simplified terms, it was thought that for a new life, it was necessary not only to provide the dead with shelter and clothing, but also with their bodily shell, so that the *vital forces* could return to the body to which they had once belonged.

No one knows who invented embalming, long experimentation with the corpses of slaves and prisoners may have preceded the refined technique. A mummification of human body can result in different ways. Pickling or embalming can be done in vinegar, wine or stronger spirits. Palm wine was used in Egyptian period. A body can be treated with silicate of potash and then inserted in mild acid solution. In another method the blood is drained off and the body is filled with formaline like preservative. Certain other examples of chemical preservation are available in history. The body of Lord Nelson was kept in brandy. When Alexander the Great died, his body was brought to Macedonia from Babylon in honey. The body of Soviet chief Lenin is kept chemically preserved in Kremlin at Red squire.

Egyptians felt that heart or *Ib* was the source of intelligence, memory and wisdom. It was also considered to be associated with bravery, sadness and love so they tried to keep the heart intact to body. Other parts were removed or chemically treated and kept in ornamental urns, called the canopic jars, which were placed in the tomb with the mummy.

Mummyin Baroda Museum

In Baroda museum and picture gallery the mummy is placed in Egypto-Babylonian gallery in a glass chamber of the size of 214 × 91.5 × 81.5 cm. Its top is having a wooden dome of 116 cm in height. The size of mummy kept in Baroda museum is 154 × 35 × 17 cm. It has a coffin box made up of wood obtained from Sekamore tree belonging to family Asclepiadaceae. This mummy belongs to late new empire of the Alexandrian period. At that time Ptollemy II and his wife, his full sister were the co-rulers. Although oldest known mummies date back to 2500BC the period of this mummy has been ascertained to 230BC by an Egyptian mummy expert Mr. Nasry Iskander, during his visit to Baroda.

Studies on Indoor Aeromycoflora of Museum

Air samples of Egypto-Babilonian gallery taken on 22nd September 1999. It is evident from the Table 1 that 25 species of fungi belonging to 15 genera were isolated. Occurrence of non-sporulating white and black mycelial fungi were also recorded. Isolations from exposed petriplates (for 10 min.) resulted in pure cultures of 7 *Aspergillus* spp. and 2 *Penicillium* spp. These biodeteriogens have the potential to cause plant and animal diseases as well as biodeterioration of cultural property. Earlier studies undertaken by Shah *et al.* (1996) revealed the presence of 22 species belonging to 12 different genera in Baroda museum. According to Nair (1981), " fungi that can grow on any suitable substrata play a prominent ole in deterioration of museum materials." The minimum and maximum relative humidity on that day was 18 per cent and 63 per cent respectively and the room temperature at the time of sampling (1 p.m.) was 32°C.

Air samples were taken on 16th April 1999 from the mummy chamber. Isolations made from colonies developed in the petridishes revealed the dominance of *Aspergillus flavus*. Other Aspergilli included *A. fumigatus* and *A. niger*. Fungal flora of the dust accumulated in the coffin box was also analyzed by dilution plate method. It showed the presence of 6 different species. Along the 4 fungi listed above *Penicillium citrinum* and *Mucor* sp. were also recorded. The dust also included dead remains of insects. The powdered wood was accumulated in the form of dust in coffin box. Various slits present in the glass chamber might have allowed the suspended dust particles to gain entry, which ultimately have settled at the bottom with powdered wood.

Table 2: Percentage Occurrence of Different Fungi in Air and Dust of Mummy Chamber on 16th April 1999

Sl.No.	*Fungi Isolated*	*Mummy Chamber*			
		Air		*Dust*	
		*Colonies**	*%*	*Colonies**	*%*
1.	*Aspergillus flavus*	10	35.7	8	29.6
2.	*A. fumigatus*	5	17.8	2	7.4
3.	*A. niger*	5	17.8	5	18.5
4.	*Emericella nivea*	2	7.1	5	18.5
5.	*Penicillium citrinum*	2	7.1	2	7.4
6.	*Mucor hiemalis*	2	7.1	5	18.5
7.	Unidentified white mycelium	2	—	—	
	Total no. of colonies	28		27	

* The data represents the total nos. of colonies appeared on 5 PDA plates after 7 days of incubation.

% = Percentage occurrence, – = Absence of colony

Aspergillus and *Cladosporium* are two universal biopollutants (Chile 1993). In a survey of atmosphere of Jabalpur 66 per cent of the trapped spores were *Aspergillus* type (Verma and George, 1997). In a survey of aeromycoflora, Khandelwal (1991) and Wadhwani *et al.* (1986) reported 20 spp. of *Aspergillus,* of these 12 were potential allergens present in the air of Lucknow. Agrawal and Dhawan (1986) reported common fungi causing damage to museum objects. Of these *Aspergillus* and *Penicillium* are the two most serious ones. As some of the species can survive for 30-40 years. Species of *Aspergillus* and *Chaetomium* have been reported earlier causing biodeterioration in Baroda museum. Arya *et al.* (1988) found association of *A. versicolor* and *A. terreus* with miniature paintings kept in Baroda museum. *A. tamarii* was found associated with miniature painting kept in museum of Department of Museology (Shah *et al.*, 1992).

Cooke and Rayner (1984) have described the ecology of saprophytic Aspergilli. According to them, "Water is required by fungi not only as metabolite and solvent but also to maintain sufficient hydrostatic pressure within hyphae to drive their apical extension". Unlike field fungi which require water potential to be above –150 bar, the storage fungi are active between – 500 and –150 bar. The presence of *A. flavus* in the aerospora of mummy chamber may be explained by prevailing hot conditions and water potential exceeding – 220 bar. Presence of fungi like *Chaetomium atrobrunneum, Lewia infectoria* and *Monascus rubur* in indoor air is new record.

Biodeterioration of Mummy

The mummy is a chemically treated dead body enclosed in linen bandages. In order to solve the arousing curiosity of the visitors, the bandage was partly removed from the feet of the mummy by museum curators. The fingers and toes are thus exposed to air. An unusual white coloured spot was observed on the 1st finger (near the toe) of the right leg. After some time an almost similar but smaller spot was noticed on the 3rd finger. To confirm the involvement of any fungus the isolations were made from these spots. Scrapings of white growth were transferred aseptically in culture tubes. Cellophane tape method was also tried to isolate the fungal biodeteriogens associated with mummy. Isolated fungus was purified and morphological characters were studied. On the basis of morphological characters, colour and reproductive spores the fungus is identified as *Emericella nivea* Wiley

& Simmons. Identity of the fungus is confirmed by Agharkar Research Institute, Pune.

Emericella Nivea Wiley & Simmons

Colonies were slow growing on PDA slants. Conidiophores measure 600 µm to 1000 µm long, heads with white globose vesicle 20 µm to 30 µm in dia. Phialides in two series, primary sterigmata 5-8 µm × 2.5-3 µm, secondary sterigmata 5- 7 µm × 2-2.5 µm. Conidia colourless 2.2-3.3 µm in size, globose, smooth. Old culture tubes containing PDA slants showed the presence of perfect state. Rounded black cliestothecia were seen in culture. These were 350-400 µm in dia. Asci were evanescent, ascospores smooth walled with a pulley, 4.4-5.5 µm in dia.

Presence of *E. nivea* in indoor air and on mummy kept in Baroda museum is new to the Science. Earlier the anamorph of this fungus *Aspergillus niveus* Blochwitz was isolated from air in northern Italy at 1,700 meters (Raper & Fennel, 1965). Studies have shown presence of another species of *Emericella* i.e. *E. nidulans* (Eldam) Vuill. in indoor air of different galleries of Baroda museum.

Acknowledgements

The authors are grateful to Prof. G. P. Senan, Head Department of Botany for providing necessary laboratory facilities. Thanks are also due to the Director, Agharkar Research Institute, Pune, and to the Head, Division of Plant pathology, IARI, New Delhi, for confirming the identity of the fungi. Assistance provided by the staff of the Baroda Museum during the study and by Shri Ravi Dhindorkar for the photography is highly appreciated.

References

Agrawal, O.P. and Dhawan S. *Control of Biodeterioration in Museums*. Technical Note: 2 N.R.L.C., Lucknow, 16p.

Arya, A., Bhowmik, S. K. and Shah, A.R. (1988). Biodeterioration of Paintings. *Geobios. New Reports*. 7: 179-180p.

Booth, C. (1971). *Methods in Microbiology.* Acad. Press, London.

Chanda, S. (1992). Aerobiology: An Inter and Multidisciplinary Approach. *Indian J. Aerobiol.* sp. Vol. 1-10p.

Chile, S. (1993). *Study of Airborne Fungal Spores at Jabalpur with Special Reference to Allergenicity.* Ph.D. Thesis, R. D. Univ. Jabalpur.

Cooke, R C. and Rayner, A.D.M. (1984). *Ecology of Saprotrophic Fungi.* Publ. by Longman London, 415p.

Gregory, P.H. (19730. *Microbiology of the Atmosphere* (2nd ed.) Leonard Hill Aylesbury, U.K. 377p.

Hirst, J.M. (1952). An Automatic Volumetric Spore Trap. *Ann. Appl. Bioi.*, 39: 257-165p.

Hyde, H.A. and Williams D. A. (1961). *Advancement of Sciences*, London, 17: 525p.

Khandelwal, A. (1991). Pollen and Spore Rain in Lucknow City during 1983-1986, In: *Biodeterioration of Cultural Property* (eds.) Agrawal, O.P. and Dhawan, S. Pub., by Macmillan India Ltd. 386-396p.

Knox, R. B. (1979). *Pollen and Allergy.* Arnold, London.

Nair, S.M. (1981). In: *Cultural Contours of India* (ed.) Vijay S. Srivastava, Abhinav Publications, Jaipur.

Nigam, R. K. and Pathak, N. C. (1991) Environmental Mycology of a Tannery and a Textile Unit at Kanpur: Allergical Aspects. In: *Biodeterioration of cultural Property* (eds.) Agrawal, O. P. and Dhawan, S. Pub. by Macmillan India Ltd. 375-386p.

Raper, E.B. and Fenell D.I. (1965). *The Genus Aspergillus.* The Williams & Willkins Company, Baltimore, U.S.A. 686p.

Shah, N. R., Shah, A. R., Arya, A. and Bhowmik, S. K. (1994). Occurrence of Fungi and Problems of Biodeterioration in the Museum and Picture Gallery, Baroda. In: *Advances in Plant Sciences* (ed) Sahni, K.C. Int. Book Distributors, Dehradun 25-46p

Shah, N.R., Arya, A., Shah, A.R. and Bhowmik, S. K: (1996). Study of Atmospheric Fungal Flora and Biodeterioration in Different Museums of Baroda. In: *Perspective in Bioi. Sci.* (eds) Rai, V. Naik, M .L. and Manoharachary, C. School of Life Sciences, Raipur,189-196p.

Shah, R.P., Shah, N.R., Shah, A.R., and Arya, A. (1992). Biodeterioration of paintings and wooden plaque by *Aspergillus* and *Cladosporium. Geobios New Reports II*: 171-172p.

Spieksma, F.T.M. (1980). Importance of Aerobiological Studies for the Prediction of Human Diseases. *Aerobiol. Symp. Munich*, 307-315p.

Thorwald, J. (1962). *Science and Secrets of Early Medicine.* Pub by Droemershe Verlanganstalt, Munich & Thames Hudson Ltd. London 331p.

Timonin, M. I. (1941). The Interaction of Higher Plants and Soil Microorganisms III Effects of by Products of Plant Growth on Activity of Fungi and Actinomycetes. *Soil Sci.* 52: 395-413p.

Verma, K. S. and George, A. M. (1997). An Assessment of the Fungal Spores in the Atmosphere of Jabalpur. *Res. J. of Chemistry and Environment* 1: 29-31p.

Wadhwani, K, Srivastava, A. K., Jamil, Z., Rizvi, H.I., Misra, A.K. and Nagi, H. (1986). Allergenic Aspergilli and Smut Spores From the Air in Lucknow. *Jour. Rec. Adv. Appl. Sci.* (1): 8-12p.

Chapter 12

Two Unrecorded Biopollutants Causing Damage to Ornamental Plants in Vadodara

Chitra Arya and Arun Arya

Botany Department, Faculty of Science,
The M.S. University of Baroda, Vadodara

Abstract

Fungi constitute major share of biopollutants present in our atmosphere. Usually they reproduce by sexual and asexual means and their spores as well as conidia form important component of aerospora. Fungal spores may cause diseases and biodeterioration of organic materials. Species of *Aspergillus* are the prime etiologic agents in allergic broncho-pulmonary aspergillosis.

Leaves have been shown to be colonized by fungi from first unfolding of the bud. The leaf surface can act as landing stage for spores and other propagules in the air, which may be deposited by impaction, by boundary line exchange, by sedimentation under the influence of gravity; and in rain and splash droplets. Presence of pollen and nutrient status of leaf plays a significant role in establishment of phylloplane micro-organisms. Presence of saprophytes affects the development of pathogens, which are directly competing for space and nutrients and sometimes showing antagonistic effect.

The present research paper reports two potential plant pathogens *Colletotrichum* state of *Glomerella cingulata* (Stonem.) Spauld. & Shrenk and *Nigrospora sphaerica* (Sacc.) Mason causing threat to two ornamental plants *i.e. Nyctanthes arbor-tristis* L. and *Tecoma stans* (L.) HB. & K. respectively. Symptoms produced by these pathogens are described and morphological characters are recorded. Pathogenicity tests were conducted. These biopollutants are new host record for the above plants.

Introduction

Fungi, bacteria, actinomycetes and pollen grains are important biopollutants. Their presence in environment may pose threat to plants, animals and human beings. A large number of aerobiologists such as Agrawal *et al.* (1969), Benninghoff (1991), Chanda (1992), Gregory (1973), Knox (1973), Nilsson and Berggren (1991), Shah *et al.* (1996) and Spieksma (1980) have emphasized the importance of monitoring such biopollutants in different localities. The shape and size of airborne particles are of interest to many specialists. Gregory (1973) recorded spread of fungi and bacteria by the action of wind and rain splash. In the past phytopathologists could identify spore stages and predict infection time and probable damage in certain areas (Stakman and Christensen 1946). The success in case of cereal rust infection patterns attracted wide attention. In tropical and sub tropical climate diseased plants can act as sources of airborne biopollution. These propagules may cause plant diseases and even allergy to human beings. The ability of a pathogen to infect the host and not the non-host continues to be enigmatic. The freedom of non-host from infection could be attributed, besides others, to its cell walls, especially in view of switching points noted during the invasion of non-hosts (Heath 1974). A fungus may fail to cause infection or show its reduced activity in spite of formation of infection structures (Politis, 1976). The influence of cell walls on synthesis of polysaccharides by pathogens or their activity may play an important role in determining resistance to invasion.

The leaf surface can act as landing stage for spores and other propagules in the air, which may be deposited by impaction, by boundary layer exchange, by sedimentation under the influence of gravity, and in rain and splash droplets. Once on the leaf the presence of pollen grains and diffusion of nutrients benefit the spores. The leaves have been shown to colonized by microbes from the first unfolding of the bud. The phylloplane micro flora may compete for nutrients and space, the organisms may show antagonism. Further, the natural pollutants like smoke dust and pollen grains of the higher plants contribute a lot to the leaf surface niche. If the microbial equilibrium on plant surface is broken or weakened the plant gets diseased.

Nyctanthes arbor–tristis commonly known as Harsingar or Parijat is a small tree belonging to Oleaceae. The leaves are medicinally important as laxative, diphoretic and diuretic. Flowers have good aromatic fragrance. They yield orange dye, which is used to colour silk. Another ornamental selected for the study is a prominent yellow flower bearing shrub or small tree. *T. stans* is a member of family Bignoniaceae. The tree is grown for beautiful flowers, which are produced, in terminal panicles. The wood is hard and durable. Roots are used medicinally and are diuretic. The studies were conducted to study the association of fungal pollutants with these two plants. Disease symptoms and other details are reported here.

Materials and Methods

During a survey of foliicolous fungi of ornamental plants of Baroda, the pathogen responsible for leaf spot diseases of *Nyctanthes* and *Tecoma* were isolated, purified and maintained on PDA slants. The fungi were isolated from the plants growing in Science Faculty campus, M. S. University of Baroda. Fungi were stained with lactophenol cotton blue (Staples, 1973). Morphological characters of two fungi were studied and identification was confirmed by Division of Pathology, IARI, New Delhi. The pathogenicity tests were conducted and Koch's postulates were fully satisfied.

Results and Discussion

Colletotrichum Leaf Spot of *Nyctanthes*

A severe leaf spot of *Nyctanthes* was observed in the months of April-May 1998 in the plants growing near Population Research Center, M. S. University of Baroda. The disease started from the tips or margins of leaves and progressed towards midrib. In severe cases the leaves became yellow and got detached. The spots were dark brown, amphigenous. Isolations from the infected leaves gave a well sporulating culture of *Colletotrichum* state of *Glomerella cingulata* (Stonem.) Spauld. & Shrenk. Experiments related to pathogenicity tests gave positive results.

The fungus *Colletotrichum gloeosporioides* Penz. is an anamorph of *G. cingulata*. A member of Coelomycetes in Deuteromycotina, it is characterized by presence of cup shaped fruiting bodies called acervuli. These bear conidiophores on which sickle shaped conidia are produced. There are presence of dark coloured septate setae. Conidia are small and hyaline (Plate 1, Photo 2). The culture has been deposited in Indian Type Culture Collection, IARI, New Delhi.

The fungus is a necrotroph. It causes leaf spot and fruit rot diseases commonly referred as *Anthracnose*. This biopollutant caused leaf spots in *Nephalium litchi* (Prasad, 1962), *Peperomia tithymaloides, Aglaonema simplex, Jasminum sambac*, and *Magnolia grandijlora* (Saxena and Maleki, 1959), *Bougainvillaea glabra* (Srivastava and Bilgrami, 1963) *Ficus benghalensis* (Tilak and Rao, 1968) *Polyalthia longifolia* (Peethambaran and Wilson, 1970), and *Santalum album* (Rao and Anahosur, 1971).

Nigrospora Leaf Spot of *Tecoma*

The disease appeared in months of January to March 1997. Browning of the leaflets was the common symptom. It started along the margins and usually from the tip of the leaflets. At later stages more than half area was covered. Leaf spots were yellowish brown in colour and amphigenous. When a major portion of the leaf had been spotted it got separated from the plant. The disease imparts an ugly look to the tree. Repeated isolation from such leaflets gave a well sporulating culture of *Nigrospora sphaerica* (Sacc.) Mason. The identity of the pathogen has been confirmed by Indian Type Culture Collection IARI, New Delhi. The pathogenicity tests were conducted and Koch's postulates were confirmed.

This biopollutant belongs to the Hyphomycetes group of Deuteromycotina. The fungus is characterised by dark coloured branched and septate mycelium. Black coloured rounded or spherical conidia are attached to hyaline vesicle. The size of conidia is 6.6 µm in dia.

The fungus *N. sphaerica* has been reported from air of Waltair and soil of Lucknow and Kanpur (Bilgrami *et al.*, 1979). It is found pathogenic on the leaves of *Ficus krishnae* (Bilgrami, 1979), *Andropogon sorghum* (Lal and Yadav, 1965), *Oryza sativa* (Pavgi *et al.*,1966), *Cajanus indicus* (Chowdhury, 1967), *Sorghum vulgare* (Chowdhury, 1969).

These two biopollutants are important pathogens of a number of economic plants. Their occurrence on two medicinal and ornamental plants is new to the science.

Photo. 1: A twig of *Nyctanthes arbor-tristis* showing fruits and infected leaves.

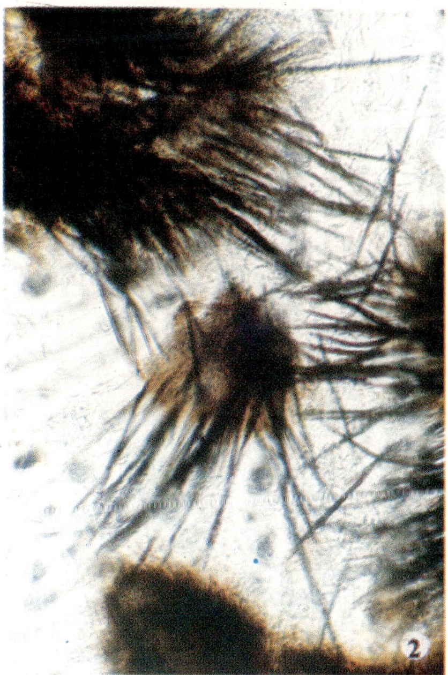

Photo. 2: Photomicrograph showing acervuli of asexual stage of *Glomerella cingulata* (note the long black setae and small, hyaline, sickle shaped conidia). X126

Photo. 3: A leaf of *Tecoma stans* showing infected leaflets.

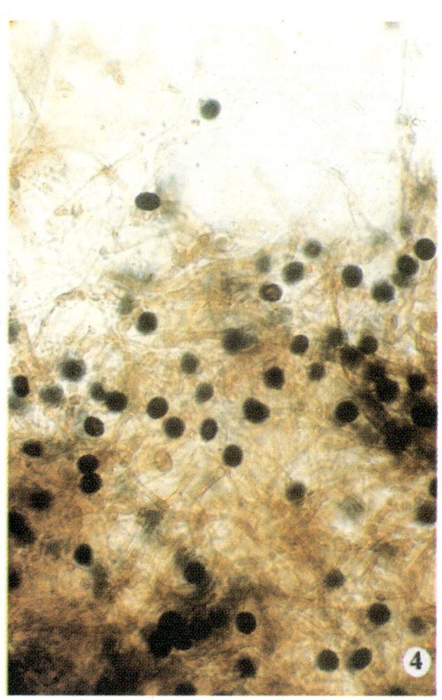

Photo. 4: Photomicrograph showing mycelium and black conidia of *Nigrospora sphaerica*. X126

Acknowledgements

We thank Head, Div. of Plant Pathology, IARI, New Delhi, for confirming the identity of pathogens and are grateful to Prof G. P. Senan, Head Dept. of Botany, M. S. University of Baroda for providing the necessary laboratory facilities.

References

Agrawal, M. K., Shivpuri D. N., and Mukerji K. G. (1969). Studies on the Allergenic Fungal Spores of Delhi India Metropolitan Area: Botanical Aspects. *J. Allergy*. 44:193-203p.

Benninghoff, W. S. (1991). Aerobiology and its Significance to Biogeography and Ecology. *Grana* 30: 9-13p.

Bilgrami, K.S (1979). Leaf Spot Diseases of Some Common Ornamental Plants. *Proc. Nat. Acad. Sci. India*. 33:429-452p.

Bilgrami, K. S. Jamaluddin and Rizvi M.A. (1979). *Fungi of India Vol. 1* Pub. Today and Tomorrow's Printers and Publishers, New Delhi 467p.

Chowdhury, S. R. (1967). Notes on Fungi Occurring at Raipur, M.P. II *Proc. Nat. Acad Sci.* India. 37B : 139-147p.

Chowdhury, S. R. (1969). Additions to Fungi of Raipur, M. P. *Sydowia* 23:46-53p.

Gregory, P.H. (1973). *Microbiology of the Atmosphere* (2nd ed.) Leonard Hill Books Aylesbury, Bucks U.K. 377p.

Heath, M.C. (1974). Light and Electron Microscope Studies of the Interaction of Host Plants and Non Host Plants with Cowpea Rust *Uromyces phaseoli var. vignae. Physiol. Plant Pathol.* 4: 403-414p.

Knox, R.B. (1979). *Pollen and Allergy*. Arnold, London.

Lall, S.P. and Yadav A. S. (1965). Additions to the Micro Fungi of Bihar III. *J. Ind. Bot. Soc.* 8: 402-406p.

Nilsson, S. and Berggren B. (1991). Various Method to Determine Air Pollutant on Pollen Grain. *Grana*. 30: 353-356p.

Pavgi, M. S. and Singh R. A. and Ramudar. (1966). Some Parasitic Fungi on Rice from India II. *Mcopath et Mycol. Appl.* 30: 314-322p.

Peethambaran, C. K. and Wilson K. I. (1970). Fungal Flora of Kerala III. *Agri. Res. J Kerala* 8: 126-128p.

Politis, D. J. (1976). Ultra Structure of Penetration by *Colletotrichum graminicola* of Highly Resistant Oat Leaves. *Physiol. Plant Pathol.* 8: 117 –122p.

Prasad, S.S. (1962). Two New Leaf Spot Diseases of *Nephelium litchi* Carob. *Curr. Sci.* 31: 293p.

Rao, V. G. and Anahosur K. H. (1971). Some Interesting Fungi Imperfecti from India. *Sydowia* 25: 51-53p.

Saxena, B. N. and Maleki Z. (1959). Leaf-infecting Fungi of Ornamental and Allied Plants of Uttar Pradesh, New Records and Host Index. *J Sci. Res.* Lucknow Univ. 1: 11-22p.

Shah, N. R., Arya A., Shah, A. R. and Bhowmik, S. K. (1996). A Study of Atmospheric Fungal Flora and Biodeterioration in Different Museums of Baroda. *Persp. Biol. Sci.* eds. Rai, V. Naik M.L. and Manoharachari, C., School of Life Sciences, Raipur. 189-196p.

Srivastava, H. P. and Bilgrami K. S. (1963). Occurrence of the Perfect Stage of *Colletotrichum gloeosporioides* Penz. on the Leaves of *Bougainvillaea glabra* Choisy. *Die. Naturwissenschaften* 50: 451p.

Spieksma, F. T. M. (1980). Importance of Aerobiological Studies for the Prediction of Human Diseases. *Aerobiol. Symp. Munich.* 307-315p.

Stakman, E. C. and Christensen C. M. (1946). Aerobiology in relation to plant disease. *Biol. Rev.* 12: 205-53p.

Staples, D.G. (1973). *An Introduction to Microbiology.* Pub. Macmillan Education Ltd. 193p.

Tilak, S. T. and Rao, R. (1968). *Second Supplement to Fungi of India* (1962-67).

CHAPTER 13

Survey of Air Pollutants from Combustion Process of Various Industries

N.M. Bhatt, Arti D. Galgale & Sangeeta Chawda

Civil Engineering Department,

Faculty of Technology and Engineering,

M.S. University of Baroda, Vadodara - 390002

Abstract

Survey of seven industries involved in manufacturing of various products was carried out. The pollutants considered were SPM, SO_2, and NO_x. The comparison of these pollutants with respect to the fuel they are using in boiler, furnace, incinerator and D G Set was done. Type of fuel used has impact on the pollutants emitted from the stack and also in ambient air.

Introduction

Fuel combustion emit mainly SO_2, NO_X, particulate, acids and aldehydes. CO a toxic contaminant is also emitted from combustion of all the fuels used in stationary combustion. Natural gas is reported to be the best as it only contains traces of sulphur and the particulate emission is also much less.

Industrial processes like metallurgical industries, chemical plants, pharmaceutical industries, Petroleum refineries, Sugar mills, pulp and Paper mills, Dyes and Dyes intermediate and synthetic rubber manufacturing plants are the major sources which contribute to about 20% of air pollution. The major sources of air pollutants are the gaseous matter and particulate matter released by burning of fossil fuels such as coal, wood, oil, etc.

The primary particulate matter consists of dust or smoke particles. This particulate matter ranges in size from 0.001 μm to several hundred μm. Man made activities emit about 450 millions tons of particulate every year, particularly dust during construction, fly ash from power plants, smelters, mining processes and smoke from incomplete combustion in stationary combustion sources of coal, wood, fuel, oil etc. miscellaneous sources include coal, refuse etc.

Control equipment generally used are:

(i) Gravity setting chamber
(ii) Cyclone collector
(iii) Filters
(iv) Scrubber
(v) Electrostatic precipitator

Sulphur dioxide (SO_2) is the second most important contributor of air pollution as it accounts for 29% of the total weight of all pollutants.

Among man made sources, fossil fuel combustion accounts for 74%, industries 24% and transportation 2% of the total SO_X emissions. This clearly indicates that coal fired power stations are mainly responsible for SO_X pollution followed by industrial plants. SO_2 can be controlled by (i) Use of low sulphur fuel (ii) using other energy sources (iii) removal of sulphur from fuel before burning (iv) removal of SO_2 from fuel gases.

Third pollutant under consideration is NO_X. Major man made activities include combustion of coal, oil, natural gas and gasoline etc. Which produce upto 50 ppm of nitrogen. NO_X are also produced as by products of some chemical industries like nitric acid and sulphuric acid industry, manufacturing of nylon intermediates etc. Another potentially serious source of man made stratospheric NO_X involves rising fibre balls associated with atmospheric explosion. Abundant fluxes of nuclear produced NO_X are immediately transported to high altitudes by aerodynamic forces created by explosion itself.

The approach for control of NOx is generally by modification in heating systems like (I) decreasing the flame temperature by injecting (i) recalculated fuel gases, water, steam (ii) Two step combustion process, (iii) catalytic decomposition.

Here we have carried out survey of various industries using different types of fuels. The measurement of suspended particulate matter (SPM), sulphur dioxide (SO_2) and oxides of nitrogen (NO_x) were done, at the source and ambient and compared with the standards.

Results & Discussion

The various industries are using fuels as LSHS, Furnace Oil, HSD, Coal & Natural Gas.

Generally these fuels are used in furnace, Boiler, D.G. Set and incinerator. Table 1 shows concentration of SPM, SO_2 and NO_x at the source and in ambient. It was observed that in steel product industry, SPM are very high at the order of 443 to 480 Hg/m^3 in ambient air while for source it is within limit. The fuel used is L.S.H.S.

For bulk drug pharmaceutical industry, the stack shows high concentration of SPM & SO_2 while in ambient air it is within the limit. In this case, the fuel used was furnace oil.

In case of Agrochemical, SPM and SO_2 concentration is exceeding the limit at boiler stack and at incinerator, SPM is as high as 278.4 mg / Nm^3 and at diesel set where the fuel is HSD, SPM is 218.68 mg/Nm^3 and SO_2 at both the places stack as well as ambient 124.6 ppm. Glue and Gelatine manufacturing industry is using L.S.H.S. as a fuel showing high concentration of SPM and SO_2 at both the places i.e. stack as well as ambient air, particularly in ambient air SPM is very high i.e. 8320 µg/m^3.

So, from the survey it was observed that in four industries out of seven SO_2 is exceeding the limit. Only at one place scrubber is installed to control SO_2 emissions from incinerator. At other four industries where SO_2 is high at stack either they have no controlling or devices are not working efficiently. In the case of SPM, at stack four industries show high concentration of SPM, again improper control of particulate matter. Also two industries have shown high concentration of SPM in ambient and out of that one industry has concentration of SPM as 8320 mg/m^3. Looking to the above scenario

Table 1

Sl.No.	Product	Combustion Unit, fuel used and control measure, if any	Concentration of Pollutants in stack gases			Concentration of Pollutants In ambient air in µg/m_3		
			SPM mg/Nm^3	SO_2 ppm	NO_X ppm	SPM	SO_2	NO_X
1.	Steel product	1. Furnace – 1 – LSHS	90	0.86	1.61	479.0	12.7	4.95
		2. Furnace – 1 – LSHS	86	0.85	0.51	443.0	76.4	42.75
2.	Bulk drug pharmaceutical intermediates	3. Boiler – 1 – Furnace oil	192	416	0.55	332	44	35
3.	Pharmaceutical	1. Boiler – Natural gas	94.17	3.98	0.92	69.65	23.02	13.20
4.	Pesticides	1. Boiler – Furnace oil	110	29.0	3.1	192	16	4.2
		2. Incinerator – Furnace oil	145	24.0	5.1			
5.	Agro chemicals	1. Boiler – fuel oil	190.5	207.7	0.42	282.2	77.8	8.4
		2. Incinerator – LSHS, alkaline water scrubber	278.4	37.5	0.96			
		3. D.U. Set – HSD	218.68	124.6	21.86			
6.	Glue, Gelatine	1. Boiler – Furnace	230	547.68	1.02	8320	517.4	2.15
7.	Textile Industry	1. Boiler – Coal	225	70	40	175	15	10

it was observed that control of particularly SO_2 and SPM is not in a proper shape.

Glue and gelatine manufacturing industry is using L.S.H.S. as a fuel showing high concentration of SPM and SO_2 at both the places i.e. stack as well as ambient air, particularly in ambient air SPM is very high i.e. 8320 μg/m³.

Typical composition of fuels has been given in Table 2. It has been observed that change in fuel is very essential particularly for SO_2 emissions. In all cases chemical and physical analysis is not carried out which is important. By analysing the particulate for physical and chemical characteristics, their effects can be predicted. As less than 10 μm size came under the category of repairable SPM. Chemical characteristics of SPM depend upon the industrial processes, and environment around it. Inorganic particulate which generate from metallic oxides are sulphides, carbonates etc. These are the products when fuel containing metals are used. Particulate of $Fe_3 O_4$ is formed during combustion of pyrite containing coal.

$$3FeS_2 + SO_2 \longrightarrow Fe_3O_4 + 6SO_2$$

Calcium Carbonate as fraction of coal converted to CaO, organic vanadium in residual fuel oil is converted to vanadium oxide V_2O_3. It is well known that lead halides are generated by combustion of leaded gasoline. $CaSO_4$, $(NH_4)_2SO_4$ and other sulphate salts are also common particulate due to SO_2.

Organic Particulate Matter (OPM) originate mainly from combustion of fuel, automobiles, vegetation and industrial process, also form using solvents and producing solvents. Polycyclic aromatic hydrocarbons (PAH) like chrysene, benzo chloroethan, benzoid pyrene, benedine etc. These are carcinogenic in nature. PAH compounds mostly occurring in urban industrial area.

Typical OPM obtained in USA is with the molecular $C_{32.8}$ H_{48} $O_{3.8}$ $S_{0.083}$ $Halogen_{0.065}$ $Alkoxy_{0.12}$. Such particles pose a serious health hazard and lie in the range of 1 μm/m size. Aldehydes, ketone, peroxides, epoxides, esters, quinons and lactones are found among the oxygenated neutral organic compounds. Organic acid present in OPM includes lauric, palmitic, stearic, myristic, oleic, linoleic and behenic acid.

Particulate paraffins are pyrolysed to C_{10} H_{22} which again disintegrate into the fine particles.

PAH compounds remain absorbed in soot particles. Soot itself is a highly condensed product of these compounds. A soot particle is composed of several thousand interconnected crystallite, i.e. graphitic plated, each having about 100 condensed aromatic rings. Soot particles act as a carrier containing toxic trace metals like Br, Cr, Mn, Ni, V, CD, Fe and poisonous organic such as Benzo oo-Pyrene.

Hence, the detailed analysis of SPM is very much essential to assess the health hazards.

Table 2: Typical Composition of Fuels

Fuels	*Percentage Combustion*							*NCV*
	C	*H_2*	*S*	*N*	*O_2*	*H_2O*	*Ash*	*KCAL/KG*
Wood	45.60	3.96	0.07	0.45	37.45	9.363	3.15	4400
Lignite	52.21	3.83	0.88	0.46	17.06	18.39	7.17	4700
Coal – 4000	44.03	3.10	0.32	0.82	4.77	3.84	43.13	3800
Coal – 4500	48.92	3.79	0.51	1.00	5.36	6.04	34.38	4250
Coal – 5500	59.38	3.15	0.38	1.04	4.86	6.00	25.19	5300
Husk	36.14	3.70	0.08	0.46	29.34	8.92	19.40	3100
Bagasse	40.94	4.58	0.04	0.23	36.65	10.53	7.03	3550
F.O.	84.00	11.00	3.50	-	-	1.00	0.50	9650
L.D.O.	85.50	11.50	3.00	-	-	-	-	10100
L.S.H.S.	86.70	11.80	1.00	-	-	0.50		9670
L.P.G.1	81.50	17.00	-	-	-	1.50	-	10750
N.Gas	66.00	23.00	-	-	-	1.50	-	8850

Note : 1. The above compositions can be widely depending on the mine, location, source etc.

2. Solid fuels in air dried condition.
3. LPG is assumed to be composed of (by volume) 10-40% of Propane and 70-60% of Butane Density – 2.32 KG/NM^3
4. Natural gas is assumed to be composed of (by volume) 76% of Methance, 10% of Ethance, 2% of Propane, 6% of CO_2 and 4% of N_2. Density – 0.83 Kg/Nm^3.

To reduce the organic pollutants, ionic liquid has emerged largely because; they have no measurable vapour pressure. They do not evaporate. Ionic liquids are liquid salts, not salts dissolved in a liquid but a salt that exist in liquid form at room temperature. They have positive charge and negative charge. There are dozens of various ionic liquids are developed. They are potentially good solvents. They are separable by super critical CO_2, for example, super critical CO_2 through a solution of naphthalene dissolve in ionic liquid 1, butyl - 3 -Methylinic Dazolium Hexa Fluorophosphate completely pull the naphthalene out which left all the ionic liquid behind.

The advantage has been shown from simultaneous production of Vinyle-Sulfone and H-Acid, where zero discharge has been planned on the basis of recovery of H_2SO_4, HCl, Water, Glauber salt and by products acetenenalide and sodium bi sulphate in addition SO_X and NO_X are also controlled by ventury scrubber system. SO_2 recovery in the form of Sodium bi-sulphite is carried out by the reaction with $NaOH/Na_2CO_3$.

Table 3: Standards for Stack Emissions

Pollutant	*Permissible concentration*
SPM	150 mg/Nm^3
SO_2	100 ppm
NO_X	50 ppm
HCl	20 mg/NM^3
Cl_2	9 mg/NM^3
NH_3	175 mg/Nm^3

Ambient air quality standards

Area / Pollutant	*Permissible concentration in µg/m3*			
	SPM	*SO_2*	*NO_X*	*CO*
Industrial	360	80	80	5000
Residential and mix use	140	60	60	2000
Sensitive	70	15	15	1000

Conclusion

The main source of pollutants in most of the industries is fuel. For control of SO_2, use of low sulphur fuel and natural gas will be better alternatives.

Generation of SO_2 in process stack can be reduced by efficient scrubbing system and bi-product recovery.

As SPM is exceeding the limit at many places, better control is required by available technology and periodical chemical and physical analysis should be included in monitoring system.

In medium scale, small scale industries for air pollution, control and monitoring is very essential.

References

Handbook, Technical Momento. Thermax Ltd.

The Report on Advantage of Simultaneous Chemical Production, 1999.

Sharma, B.K. & Kaur. 1997-1998. Environmental Chemistry.

Technorama Journal of the Institution of Engineers (India) 40, August 1999.

Chapter 14

Methanol Fumigation in DI Diesel Engine for Control of NO_X

K. Balagurunathan, S. Prasad, R. Ramanadhan, R. Ram Kumar and K. Thanaraj

Mechanical Engineering Department
College of Engineering, Anna University,
Guindy campus, Chennai

Abstract

This paper describes the state of art of methanol fumigation in DI diesel engine for reducing oxides of nitrogen. Formation of oxides of nitrogen and effects of NO_X are briefly described. Combustion concepts and experimental set up are dealt in detail. Fumigation technique is promising and efficient method for the control of NO_X.

Oxides of Nitrogen

Oxides of nitrogen, NO_X are also the major emissions from a diesel engine. The NO_X is the collective term used for oxides of nitrogen from a diesel engine exhaust and is mostly made of nitric oxide, NO and nitrogen dioxide, NO_2, NO is colourless and without any smell. In air, No oxidizes to NO_2 and has a pungent and irritations smell as per Miller *et al.* (1989).

The nitrogen oxides in exhaust gases are formed through reaction between nitrogen and oxygen. This reaction is affected by conditions prevailing in the combustion chamber. High temperature and pressure, along with the less residence time available for reaction. With good availability of oxygen is the cause or high NO_X emission in diesel engines.

Effect of NO_x

Ozone Depletion

The ozone layer is stratosphere serves as a shield protecting the earth's surface from most of the ultraviolet radiation in the sun's radiation. However, this layer is getting deplected by the reactions involving a variety of compounds which reach the stratosphere of particular concern are NO_x and water vapour. C.S. Rao (1991) has postulated that the effect of these would cause considerable depletion of the ozone layer and could result in an increase in ultraviolet radiation reaching the earth leading to crop damage and marked rise in the incidence or skin cancer. The recent reports of the ozone hole in the atmosphere over the Antarctica is a cause of great concern.

Combustion Concepts

Fumigation or supplementary vapour fuels into the diesel intake introduces a homogeneous fuel/air mixture into the cylinder prior to compression, the injection and ignition providing the supplementary vapour fuel has a sufficiently high auto-ignition temperature and the furnicated charge is sufficiently lean, ignition cannot occur before a robust, high energy flame front is brought into it by some other means. The diesel fuel injection and its compression-ignition phenomenon represent such means.

The existence of a lean, homogeneous fuel / air charge in the diesel combustion chamber helps overcome some of the handicaps that arise with increasing load. Since the fumigated charge in subjected to compression heating, some pre flame reaction is likely. The exotherm produced there by contributes to the charge temperature rise according to Alvin Lour (1984). Free radicals also form, increasing the chemical reactivity of the charge. These effects combine to accelerate the ignition of the injected fuel and reduce the cetane requirements of the engine. Combustion may be accelerated thereby to such a degree as to approach a constant -volume process. Under the conditions combustion can be completed there nearly in the vicinity of for-dead-center, which produces relatively higher pressure for shorter durations and permits the expansion process to occupy greater crank angle characteristics of an auto cycle. As a result, higher power output and thermal efficiency can be obtained. This also accounts in part for the usual reduction in smoke, cylinder deposits and oxides of nitrogen. While hydrocarbon and carbon-dioxide emissions tend to be increased, cyclic pressure variations are reduced, which accounts for substantial reductions in combustion -induced engine noise.

Further characteristics of inlet fumigation in diesel operation include possible improvements in air utilization and maximum shaft speed. These factors combine to add considerably to the potential for improving the diesel's smoke -limited specific power. Such results can be inferred from the otto-cycle analogy. However, the theoretical basis can be stated more specifically in terms of the effects on the combustion process. Reduced ignition delay, along with increased combustion speed, consumes proportionately less crank angle, which in turn improves the mean effective pressure of the cycle. This characteristic can be exploited for increasing peak torque, maximum power, best economy or any combination of the three.

The effect of fumigation on air-utilization is probably the most significant agony for increasing smoke-limited power output. The presence of the homogeneous fuel component reduces the concentration gradients inherent in the conventional diesel combustion chamber. This permits operation at charge-energy densities approaching the levels attained for lean best power in spark ignition engines. For example a typical smoke limited equivalence ratio for the conventional direct -injected diesel might be of the order of 0.75. With fumigation this might be increased to the level of 1.05, which represents a potential increase in torque of more than one third.

Experimental Set-Up

The experimental set up used by Qiqing Jiang *et al.* (1990) is shown in the Fig. 1. This set up is proposed to be used in our on going research on single cylinder DI diesel engine with little modification. They used for their study with John Deere 4276T four cylinder, four stroke, turbocharged diesel Engine. The bore is 106.5 mm the stroke is 127 mm, and the compression ratio is 16.8 to 1. The combustion system is a bowl-in-piston, direct injection, medium swirl type.

The Engine directly coupled to a general D.C. dynamometer, was equipped with a speed governer, which was left intact. A general Electric siltron dynamometer controller allowed speed contrcl of the engine, and the governer lever was adjusted to set engine torque.

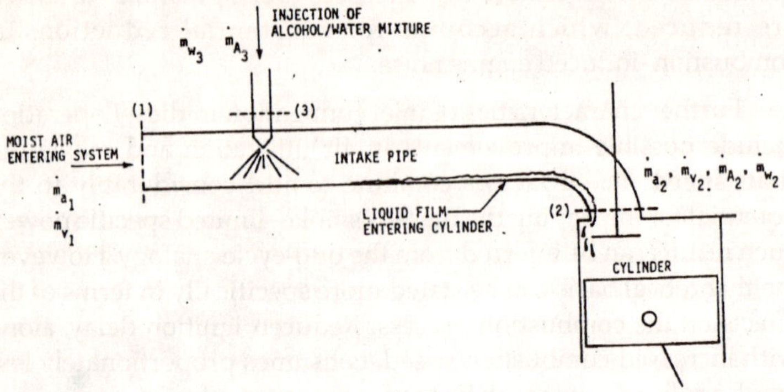

Control volumes for intake pipe and engine cylinder

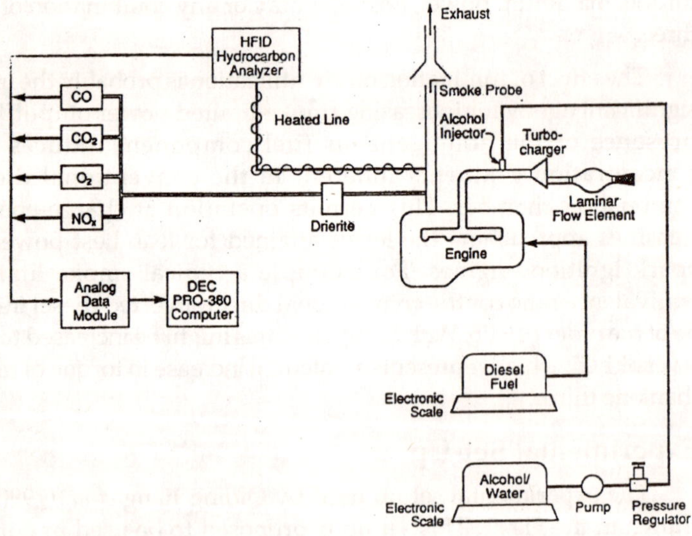

Fig. 1: Schematic of experimental apparatus

The torque was measured with a Labour strain gauge load cell mounted on the dynamometer.

Diesel fuel was supplied to the engine by use of the original. equipment transfer pump and injection pump. The alcohol and water mixture was injected into the intake manifold of the engine after the turbocharger compressor but about ten inches before the pipe branches for the individual cylinder as shown in Fig. 1. A commercial fumigation system produced by Midweth power concepts. This system includes an injector, a 12 volt electric fuel pump, a bypass pressure regulator and a positive fuel shutoff solenoid. The diesel fuel and alcohol flow rates were measured using Toledo and Sartorius electronic scales, respectively. A stop watch was used to time the mass of diesel fuel and alcohol supplied to the engine in a three to five minutes period to obtain the average flow rate. A laminar flow element was used to measure the air flow rate into the engine. Thermocouples are located at the air intake to measure the dry and wet bulb temperature of the intake air; in the exhaust ports of the individual cylinder exhaust temperatures were measured.

Exhaust emissions were sampled about 100 mm downstream of the turbocharger. An electrically heated sample line carried exhaust to a Beckman Model 402 heated flame ionization detector (HFID) type hydrocarbon analyzer. A second sample line was used for the carbon monoxide (Co), carbon dioxide (CO_2) oxides of nitrogen (NO_x) and oxygen (O_2) analyzer. This sample was filtered and dried using drierite immediately after removal from the exhaust pipe. A Beckman Model 955 chemiluminescent NO_x analyzer was used to measure oxides of nitrogen. Backman Model 864 NDIR analyzer were used for measuring carbon-dioxide and carbon mono-oxide. A Backman Model 7003 Polarigraphic oxygen analyzer was used to measure oxygen. The engine smoke was measured using a Bosch Smoke Meter. The probe of the smoke pump was inserted into the exhaust stack about 2500 mm downstream of the turbocharger.

A digital equipment corporation Analog Data Module (ADM) was used with a digital Equipment PRO-380 Computer to record data from the emission equipment. The ADM sampled each instrument once per second. The current reading was displayed on the computer monitor and an average reading for each 20 sec. period was stored for later analysis by them.

Conclusion

The state of art of methanol fumigation is described in this paper. Combined premixed-diffusion flame combustion in a fumigated diesel engine reduces flame temperature which results in reduced oxides of nitrogen in diesel engine exhaust. Cost of the existing system is increased by the cost of the extra injector and pumping system. Fumigation is one of the promising method for the reduction of oxides of nitrogen.

References

Alvin Louri, Jr. (1984). Supplementary Fuelling of Four Stroke Cycle Automotive Diesel Engine by Prepare Fumigation. SAE TRANS 841398.

James A., Miller and Craig T. Bowmem (1989). Mechanism and modelling of Nitrogen chemistry in combustion. Prog. Energy Combustions Science, Vol. 15.

Qiging Jiang, Pradheepan Ottikutti, Jon Van Gerpln and Delmar Van Meter (1990). The effect of alcohol Fumigation on Diesel Flame Temperature and emissions. SAE TRANS 900386.

Rao, C.S. (1991). Environmental Pollution Control Engineering. Wiley Eastern Ltd.

Chapter 15

Avifauna Amidst Industrial Pollution

G.S. Padate and S. Sapana
Division of Avian Biology,
Department of Zoology, Faculty of Science,
M.S. University of Baroda, Vadodara - 390 002

Abstract

Industrial development though an essential requirement for modernization, have become a major concern for the environment. Sustainable use of earth's inexhaustible resources, without sidelining the development, have been realized by man. A very good example in relation to this is set by IPCL, Vadodara unit.

An experimental ecofarm was set up in 1984, with a pond regularly filled with treated effluent, and the same was used for watering the plants. A study was conducted to find out the potential avifauna in this manmade ecosystem. 94 bird species have been listed so far. It was observed that ecofarm shelters mainly arboreal and terrestrial species, with adequate number of aquatic species in the pond which is inconsistently filled with treated effluent. Thus, a small pond, a miniforest and cultivated fields and gardens together with serene ambience provide shelter, roosting and nesting sites for a large number of species of resident and migratory species of birds.

IPCL's effort's in helping to balance the degrading environment amidst the chain of industries is worth appreciating.

Introduction

In recent years the widespread applications of science and technology is leading to problems of pollution which in turn results into depletion of earth's not in exhaustible resources with their biodiversity. As development and environment are equally essential it becomes important to maintain a steady balance between both of them. Sustainable use of the renewable resources without damaging them has become a major priority and the developing countries have started to recover, reuse and recycle all resources wherever possible. In India also, this is becoming popular and one such example has been set by the Indian Petrochemicals Corporation Limited (IPCL) at Vadodara (Baroda -22° 15' N and 73° 15'E).

Vadodara is a developing industrial area in Gujarat State, India, with several large and small industrial sectors. The IPCL is a Government of India undertaking, manufacturing and marketing a variety of petrochemical products. In 1984 IPCL developed an Eco-Experimental farm, with a small pond, filled inconsistently with treated effluent, to find out the effect of effluent on the vegetation. As much as 7 million gallons of treated effluent, released daily from the several industrial units at the petrochemical complex is channelled in a common Baroda Effluent Channel. Some treated water from this was used in raising vegetation at the ecofarm. This ecofarm is situated amidst several petrochemical units that do release some gases in the environment. This ecofarm provides varied habitats *i.e.* a small pond, agricultural fields, flowerbeds, plant nursery and a miniforest (thickest), attracting a number of birds round the year. Menon and Oza (1987) conducted a study of this ecofarm in mid 1980s and listed 78 species of birds. A decade later, a study was conducted from October 1995 to October 1998 to find out the possible ecological succession in the bird populations over a decade at this manmade ecofarm. The air in this area is having various gases released by the large petrochemical units. Though the pollution is under control, some gases are always present which can lead to acidification of soil.

Study Area

This man-made eco-farm at IPCL was developed at about 15-km. northwest of Vadodara in 1984. It spreads in an area of 14 ha. And is having a small pond of 30x80 m size, which was mainly filled with treated effluent or infrequently with fresh water in recent times. As petrochemical industrial units surround the ecofarm the air is always full of different gases. This ecofarm is expected to combat with these gases too. About 70 different plant species have been planted in specified areas of the farm (Fig.1). Seasonal cereal crops and flowers are grown here and they were watered with treated effluent. According to the 1986 report the quality of effluent is well under CWPCB standard. Large petrochemical complexes surround the ecofarm on almost all the sides, which increase the air pollution in the area.

Materials and Methods

Morning observations were made while walking on a selected transect covering different areas of the farm at least twice in a month from Oct 1995 to Oct 1998. The birds were observed using 10x 50 field binoculars and their identification was done with the help of standard books (Ali, 1979; Ali and Ripley 1983). A list of birds was prepared and then birds were divided into six different categories as aquatic, ground, arboreal, ground and arboreal, birds of prey and flying birds like swifts and swallows which we hardly see perching. This list was compared with the list prepared ten years back by Menon and Oza (1987).

Results and Discussion

Birds are considered as one of the best indicator for any obnoxious condition including pollution (Scheuhammer, 1988, 1991; Root, 1990). In the mid 1980s INSONA (International Society of Naturalists), a Vadodara based NGO had prepared a checklist of birds of this man made ecofarm when it was being established. Almost a decade later a study was conducted from Oct. 1995 to Oct. 1998 to find out the status of avifauna of this area. When these two studies are compared interesting differences have come to light with the possible immigration and emigration of birds after 10 year span have came to light.

The birds listed in the present study were categorized into 6 groups (Table 1). Seventy-six species of birds were listed, by INSONA whereas 94 are listed in the present study. Of the total 123 species listed from both the studies combined, 52 species are common in both the studies whereas 27 species were sighted only during INSONA study and 44 new bird species have been spotted during the present study. This indicates that about 27 species might have left the area and 44 new species might have taken up their vacant niche.

Table 1: Change in the Avifaunal Diversity of IPCL Experimental Ecofarm in a Decade

Habitat Study Time	*AQ*	*GR*	*ARB*	*GR & ARB*	*BOP*	*FL*	*Total*
INSONA 84-85	23	12	26	10	3	2	76
95-98	27	7	39	12	6	3	94
Common	18	5	19	7	2	1	52
Emigrants	7	7	8	3	1	1	27
Immigrants	9	2	24	2	4	3	44
Total Species	34	14	51	12	7	5	123

AQ – Aquatic Species, GR – Ground Species, ARB – Arboreal Species, GR & ARB – Species common in ground and arboreal, BOP – Birds of Prey, FL – Species seen flying.

The air pollution is known to influence bird populations in several ways like influencing egg shell quality (Eeva *et. al.*, 1995), influencing breeding activities (Morrison, 1986: Eeva and Lehikoinen, 1996; 1997) influencing food supply (Eeva *et. al.*, 1997); local survival rates (Eeva and Lehikoinen, 1998) and plumage fading (Eeva *et. al.*, 1998). Eeva and Lehikoinen (1998) include emigration also in the local survival rate where individuals leave the area in search of better habitat. Further, pollution may have direct or indirect effect on breeding behaviour of birds (Brotons *et. al.*, 1998). However, present study simply checks the avifaunal diversity that seems to have increased significantly over the time period in an experimental farm located amidst petrochemical industry. These birds can be the subordinate individuals which have immigrated to the areas of lower quality due to competition in other areas as indicated by Lundeberg *et. al.* (1981) and Alatalo *et. al.* (1985). Further, as Eeva *et. al.*, (1997) suggest the birds may be of better quality at their arrival but their condition may deteriorate under pollution stress. However no unhealthy bird was ever observed in the present study area.

The impact of untreated effluent on birds is well known (Bhanumathi and Thirumurthi, 1995; Pilo *et. al.*, 1995). However in the present study, where treated effluent was released, 31 species of aquatic birds are listed. Twentyfive species of aquatic birds like wader, ducks, kingfishers, etc. were spotted in the pond all throughout the year. Though the total number of aquatic birds species is almost the same in both the studies, it seems that 7 species probably emigrated from the ecofarm and almost an equal number of 8 new species have taken up their place. Sixteen species of aquatic birds are common to both the studies. Of the 7 possible emigrants, birds like plovers and stints require open mudflats to run about. The manmade pond in the developing Ecofarm during 1980s with scanty scrub around must have attracted these species of birds for foraging. The small sapling planted around the pond during 1980s have grown prodigiously into trees in the 1990s, which seem to have favoured the perching birds like egrets and herons presenting an example of ecological succession. As the habitat structure changes the composition of faunal diversity changes (Sufter and Bringham, 1998; Odurn, 1971) resulting into ecological succession. Many species of birds are probably trying to adapt to different habitat provided by this manmade Ecofarm. When the natural habitats are destroyed due to urbanization, deforestation and draining of wetland, there are number of hardy and able species that have come to terms with man's activities (Flegg, 1974). At present we do not know the effect of this treated effluent as well as air pollution on the physiology and biochemistry of these birds.

Another point that came into light is the significant decrease in the ground level birds. There can be two reasons for the same. Firstly, as stated earlier during mid 1980s the newly developed ecofarm probably provided open areas for these ground birds. After 10 years the open area is transformed in to a mini forest (Fig.1) attracting arboreal birds. This can be another example of ecological succession. This also explains why the numbers of arboreal bird species have increased significantly from 26 to 39 of which 24 are probably immigrants. Secondly, as reported by Graveland *et al.*, (1994) acidification decreases ground layer vegetation influencing the ground level fauna, especially snails which are rich source of Calcium and influence breeding activities. The acidification brings the pollution to ground hence, there is more accumulation of pollutants in the prey items at ground level then in foliage. Further, in polluted area ground becomes barren and well lighted because most ground vegetation is lacking. (Eeva *et al.*, 1997). However, it is interesting to note here that the Common Peafowl *Pavo cristatus* a terrestrial bird that was not listed in the INSONA study is now present in sizable breeding population. Breeding activity of *Pavo cristatus* in this area needs further investigation.

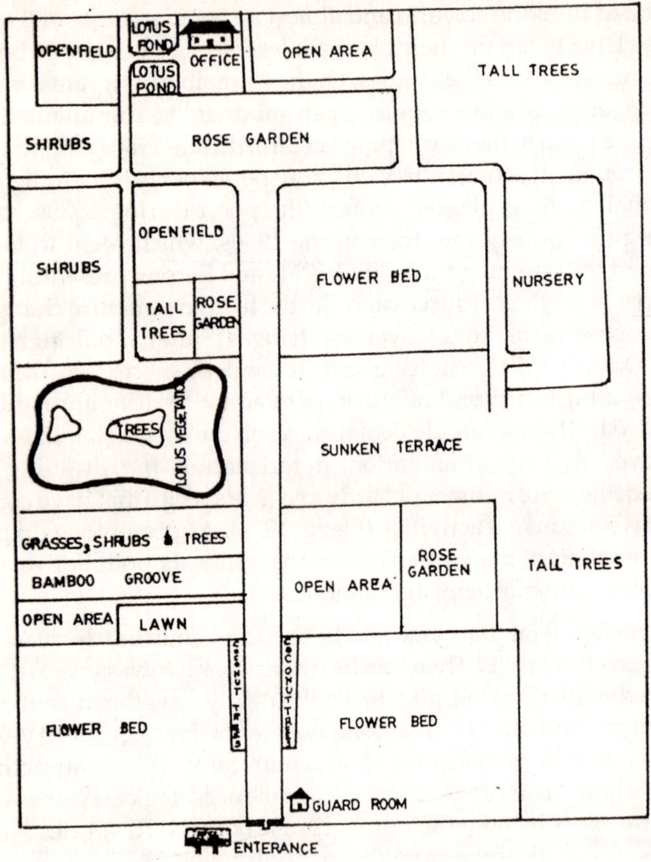

Fig. 1: Map of IPCL Eco-Farm

With the increase in the number of birds perching on the trees it is evident that the tree species planted in the ecofarm are definitely attracting arboreal birds. Small insectivorous passerines are placed at higher tropic level in food chain and hence are consider to be good , indicators for pollution too (Scheuhammer 1988, Root, 1990).

The availability of suitable prey has attracted Birds of Prey to the Ecofarm. During 1980s only three species of birds of prey were listed whereas in the present study the number has increased to six. Out of these two are common in both the studies whereas four species are new in the present study. The Kestrel which was a common bird in the past decade around Baroda city has become rare in the area and was not spotted in the present study. Four species of swifts and swallows were seen in the ecofarm twittering and twisting in air as usual.

Fifteen species of migratory birds were also spotted in the pond. Out of these 5 were waders whereas others were mainly wagtails. Besides the presence of variety of avifauna, this Ecofarm also harbours many other forms of life, like squirrels, hares, snakes and lizards, frogs and tadpoles, snails, spiders, butterflies and other insects that form the prey species of birds, forming it a balanced ecosystem.

When the presence of all the birds present in this comparatively smaller area during each month was compared it was seen that maximum birds were observed during Nov.-Dec. and April-May. The significant increase in the total number during November-December can be attributed to the visiting winter migrants *i.e.* wagtails and few waders, resident water birds like Cattle Egret *Egreta garzeta*, Red Wattled Lapwing *Vanelus indicus*, Black winged Stilt *Himantopus himantopus* and some terrestrial birds. Total number decreased during January and February when the terrestrial birds as well as some waders left. During the next couple of months (March-April) the total number increased possibly due to the return of terrestrial birds when trees started flowering. As stated earlier, these populations could be of those subordinate individuals which might have immigrated to the areas of lower quality (Lundberg *et al.*, 1981: Alatalo *et al.*, 1985) due to competition in other areas. During this time cereal crops and some other vegetables were also grown. As the breeding season (May-June-July) arrived the total number of the birds spotted was almost constant, as many of the birds had settled down for breeding. A sharp decline in the total number during August and September indicate dispersal of breeding birds from the area.

Though we do not know about what is the effect of this effluent water as well as air pollution on the birds present in this area, IPCL's efforts in keeping the environment clean is worth appreciating. In another area of 20 ha. land, a "Living Museum of trees" with 51 species of trees have been grown. Twenty nine species of birds have been listed in this "Living museum of trees" during a short study from September 1993 to February 1994 (Padate *et al.*, 1994). The present study reports presence of various species of birds, and the possible ecological succession of bird species in the area. Further investigations on quality of birds and their breeding activities are required to further appreciate IPCL's efforts in combating the pollution created by this giant petrochemical industry.

Conclusion

To improve and to maintain the ecological balance around Baroda industrial area IPCL has developed an ecofarm using treated effluent from their Units. This manmade ecofarm attracts a variety of resident and migratory birds. When compared to 1980s data it seems that though some bird species have emigrated from the area almost an equal number of bird species have occupied the vacant niche.

In this era of industrialization, urbanization and deforestation, IPCL eco-experimental farm seems to provide these bird species an artificial habitat where they can feed, rest and reproduce. Thus it can be concluded that this kind of Ecofarm can become a model for other industries wanting to preserve our environment and also control pollution.

Acknowledgement

The authors would like to thank Mr. K.Z. Parmar, Assistant Horticulturist and IPCL authorities at the Ecofarm for allowing us to carry out the study.

References

Alatalo, R. V., Lundberg, A. and Ulfstrand, S. (1985). Habitat selection in the Pied Flycatcher Ficedula hypoleuca. In: Cody, M.L. (ed.) *Habitat selection in birds*. 59-63. Academic press, London. As cited by Eeva, T. et al., 1997. Loc cit.

Ali, S. (1979). *The book of Indian Birds*. Bombay Natural History Society. Oxford University Press, Bombay.

Ali, S. and Ripley, D.S. (1983). *A pictorial Guide to the Birds of Indian subcontinent*. Bombay Natural History Society Centenary Publication. Oxford University Press. New Delhi.

Bhanumathi, C. P. and Thirumurthi, S. (1995). Impact of industrial effluent on the midwinter populations of waterfowls. Abstract presented at 2nd symposium of Ornithological Society of India, New Delhi, November 14-16, 1995.

Brotons, L., Magrans, M., Ferrus, L. and Nadal, I. (1998). Direct and indirect effects of pollution on the foraging behaviour of the forest passerines during the breeding season. *Can. J. Zool.,* 76: 556-565p.

Eeva, T. and Lehikoinen, E. (1995). Egg shell quality, clutch size and hatching success of the great tit (Parus major) and the pied flycatcher (Ficedula hypoleuca) in an air pollution gradient. *Oecologia,* 102: 312-323p.

Eeva, T. and Lehikoinen, E. (1996). Growth and mortality of nestling great tits (Parus major) and pied flycatcher (Ficedula hypoleuca) in a heavy metal pollution gradient. *Oecologia*, 103: 631-639p.

Eeva, T., Lehikoinen, E. and Pohjalainen, T. (1997). Pollution related variation in food supply and breeding success in two hole-nesting passerines. *Ecology,* 78 (4): 1120-1131

Eeva, T. and Lehikoinen, E. (1998). Local survival rates of the pied flycatchers (Ficedula hypoleuca) and the great tits (Parus major) in an air pollution gradient. *Ecosciences*. 5(1): 46-50p.

Eeva, T., Lehikoinen, E. and Ronka, M. (1998). Air pollution fades the plumage of the Great Tit. *Functional Ecology,* 12: 607-612p.

Flegg, J.J.M. (1974). *The world atlas of birds*. Mitchell Beazley Publishers Ltd. London.

Graveland, J. (1990). Effects of acid precipitation on reproduction in birds. *Experientia*, 46: 962-970p.

Graveland, J., Wab, R. van der, Balen, J. H. van, and Noordwijk, A. J. van. (1994). Poor reproduction in forest passerines from decline of snail abundance on acidified soils. *Nature*, 368: 446-448p.

Lunderbeg, A., Alatalo, R.V., Carlson, A. and Ulfstrand, S. (1981). Biometry habitat distribution and breeding success in the Pied Flycatcher Fecidula hypoleuca, Omis Scandi. 12: 68-79p. As cited by Eeva, T. *et al.* 1997 *loc cit.*

Lundberg, A., and Alatalo, R. V. (1992). The pied flycatcher. T & A D Payer. London. As cited by Eeva, T. et al 1997. *loc cit.*

Menon, G.K. and Oza, G.M. (1987). *The bird sanctuary in the IPCL complex*. INSONA study. (International Society of Naturalists, Baroda - Report).

Morrison, M.L. (1986). Bird populations as indicators of environmental change. *Current Ornithol.*, 3 : 429 -451p.

Odum, E.P. (1971). *Fundamentals of Ecology*. Harcourt Brace and Co., USA.

Padate, G.S., Barve, S.A., Garg. S., Jatinder, K., Patel, G. S., Jagirdar, N., Sapna, S. and Vani, N. (1994). Bird Species Richness as Influenced by degree of Human Intervention in Baroda. *Birds in Agricultural Ecosystem* (edited by M.S.Dhinsa, P.S.Rao and B.M.Parasharya, Society for Applied Ornithology (India), Hyderabad. 1998. pp 158-165p.

Pilo, B., Parikh, P.H. and Padate, G.S. (1995). Influence of industrial pollution on the avifauna of Mahi river ecosystem. Abstract presented at 2nd symposium of Ornithological Society of India, New Delhi, November 14-16, 1995.

Root, M. (1990). Biological monitors of pollution. *Bioscience*, 40: 83-86p.

Scheuhammer, A.M. (1988). Chronic dietary toxicity of methylmercury in the Zebrafinch Poeptila guttata. *Bull. Environ. Contam. Toxicol.*, 40: 123-130p.

Scheuhammer, A.M. (1991). Effects of acidification on the availability of toxic metals and calcium to wild birds and mammals. *Environ. Pollu.* 71: 329 - 375p.

Sutler, G.C. and Brigham, R.M. (l998). Avifaunal and habitat changes resulting from conversion of native prairie to crested wheat grass: patterns at songbird community and species levels. *Can. J. Zool.* 76 : 869-875p.

CHAPTER 16

Methane Efflux and Climate Change

S.N. Singh
Environmental Science Division,
National Botanical Research Institute, Lucknow - 226 001

Introduction

There are many greenhouse gases (CO_2, CH_4, N_2O, O_3, water vapour) present in the atmosphere in traces to provide essential warmth to the earth's surface to sustain human life. In their absence, our earth would have inhabitable due to sub-zero temperature (-18°C). These radiatively active gases reflect and absorb the IR radiations (long wave) reflected back from the earth's surface, while they are fairly transparent to incoming short wave radiations. Thus, these gases maintained the earth's surface mean temperature at +15°C. However, in recent years, increasing abundance of these gases in the atmosphere is a matter of serious concern worldwide, as it has already pushed up the earth's surface temperature, affecting the food security through changes in rain pattern. Studies have shown that the concentrations of CO_2 have increased by some 27% over the last century, while methane has increased by a factor of 2.4 and nitrous oxide has gone up by 7 per cent. These increases can be quantitatively correlated to their emissions from anthropogenic activities. The main source of CO_2 emission is burning of fossil fuels coupled with low sequestering of CO_2 due to deforestation. Increases in methane have come from intensive rice agriculture, domestic cattle and other ruminants, natural gas use, coal mining, landfills, biomass burning and a number of other smaller sources. The increase in nitrous oxide has been associated with more use of nitrogen-based fertilizers in modern agricultureand other small sources. Water vapour is also increasing, although long term measurements over large regions are not available to prove it. In addition, we are alo adding new gases, which are not the part of natural cycle, to the atmosphere. Among them, chloroflurocarbons ($C\,CI_3\,F$ and $C\,CI_2\,F_2$) are the most important exotic molecules, which contribute many times more to global warming than the naturally produced gases on molecule to molecule basis. Fortunately, their concentrations are very low to have their significant impact on the climate.

As these gases continue to increase unabated in the atmosphere, they are bound to increase the earth's surface temperature. This concern has boosted the research worldwide to upgrade the inventory of greenhouse gases to determine the source and sink strengths and to find out options to contain their emissions. In this regard, Kyoto protocol is a recent agreement among the different countries to lay down a framework for reducing greenhouse emissions. In this agreement, the gases identified for possible control are CO_2, CH_4, N_2O, SF_6, perflurocarbons (PFCs), hydroflurocarbons (HFCs) and hydro chlorofluorocarbons (HCFCs). Efforts have to be integrated at national and global levels to arrest the impending danger of greenhouse effect.

Methane and Climate Change

Although CO_2 is the main culprit for the global warming, CH_4 occupies the second position (Houghton *et al.*, 1992). CH_4 concentration in the atmosphere is very low (1.72 ppm) as compared to CO_2 (350 ppm), but its global warming potential (GWP) is very high, as it has 30 times more potential to absorb the infrared radiations reflected from the earth's surface with respect to CO_2 on molecule to molecule basis (Table 1). Methane has a short atmospheric life time of 10 years and plays an important role in photochemical reactions in the troposphere and the stratosphere and its changes do affect the chemistry of the atmosphere (Thompson and Cicerone, 1986; Khalil and Rasmussen, 1987). Although there has been an increase in CH_4 concentration over the past 200 years, a considerable decrease in the growth rate of CH_4 abundance from 1.2% per year in the late 1970s to about 0.3% per year in 1992 was noted. However, the reasons are still obscure; there may be imbalances in the source and sink strengths (Prinn, 1994; Rudolph, 1994; Bekki *et al.*, 1994).

Table 1: Characteristics of some Greenhouse Gases and their contribution to the Greenhouse effect

	CO_2	CH_4	N_2O	CFC
% share of anthropogenic GHE	50	19	6	17
Concentration (ppmv)	352	1.72	0.31	0.0002–0.0004
Residence time	100	10	170	50-150
Increase (% / year)	0.5	1.0	0.25	5.0
Relative potential for thermal absorption (CO_2=1)	1	30	150	>10000

The main sources and sinks of atmospheric CH4 are well documented (Crutzen, 1991; Houghton *et al.*, 1992). Among these sources, natural wetlands and irrigated rice fields account for about one third of the total estimated source strength of 515 Tg CH_4/yr (Houghton *et al.*, 1992). The present burden of methane in the atmosphere, which corresponds to an average mixing ratio of 1.7 ppmv in the atmosphere, is estimated to be 4800 Tg (Cicerone and Oremland, 1988). Analysis of air bubbles trapped inside polar ice cores indicates that in past 200 years, atmospheric CH_4 concentration has doubled (Rasmussen and Khalil, 1984). The present CH_4 increase is a consequence of population growth and economic development. Along with CH_4 sources, there are some natural sinks also to remove CH_4 from the atmosphere to maintain a balance. Although, the quantification of CH_4 removal by the soil has been not done, its reaction with hydroxyl radical has been effective to destroy methane in the range of 400-600 Tg/yr with a left over in the range of 40-48 Tg CH_4/yr for the atmospheric abundance (Table 2). The balance between the microbial processes of methanogenesis and CH_4 consumption controls to a great extent whether a certain terrestrial ecosystem will function as a source or sink of atmospheric methane. While methanogenesis occurs in strict anaerobic environment in which organic matter undergoes anaerobic decomposition, CH_4 is oxidized in oxic conditions. In general, process of CH_4 oxidation in soils remains less understood than those of methanogenesis (King and Adamsen, 1992; Topp, 1993).

Methane Emission

The main processes of methane transport to atmosphere are ebullition, diffusion and transport through the aerenchyma of

Table 2: Global Budget of Atmospheric Methane*

Methane	*Tg/yr*
Sources	
Enteric fermentation	65 - 100
Natural wetlands	100 - 200
Rice paddies	25 - 170
Biomass burning	20 - 80
Termites	10 - 100
Land fills	20 - 70
Oceans	5 - 20
Fresh water lakes	1 - 25
Methane hydrate destabilization	0 - 100
Coal mining	17 - 50
Gas drilling, venting, transmission	25 - 50
Total	400 - 600
Sink	
Removal by soils	?
Reaction with hydroxyl radical	400 - 600
Atmospheric increase	40 - 48

* Based on Cicerone and Oremland (1988) and IPCC (1990).

vascular plants (Seiler *et al.*, 1984; Nouchi *et al.*, 1990; Delwiche and Cicerone, 1993) (Fig. 1). Ebullition takes place whenever supersaturation of a single gas or a mixture of gases in a liquid occurs and the partial pressure of the gas exceeds the value of the hydrostatic pressure. Once bubbles are formed, they remain stable and, are able to grow larger in size until their buoyancy is enough to let them reach the surface and burst. Such increases in partial pressure might be reached especially in organic-rice sediments during warm periods from early summer to autumn, providing ideal growth conditions for the methanogenic bacteria. Secondly, dissolved CH_4 may diffuse depending on gradient concentration through the sediment-water and air-water interface. Diffusion coefficient depends on wind speed and temperature differences between air and water. Increasing wind speed decreases the boundary layer resistance, leading to higher methane emission to the atmosphere. Besides, increasing temperature affects the gas solubility and viscosity of water, both leading to gas escape to the atmosphere. In

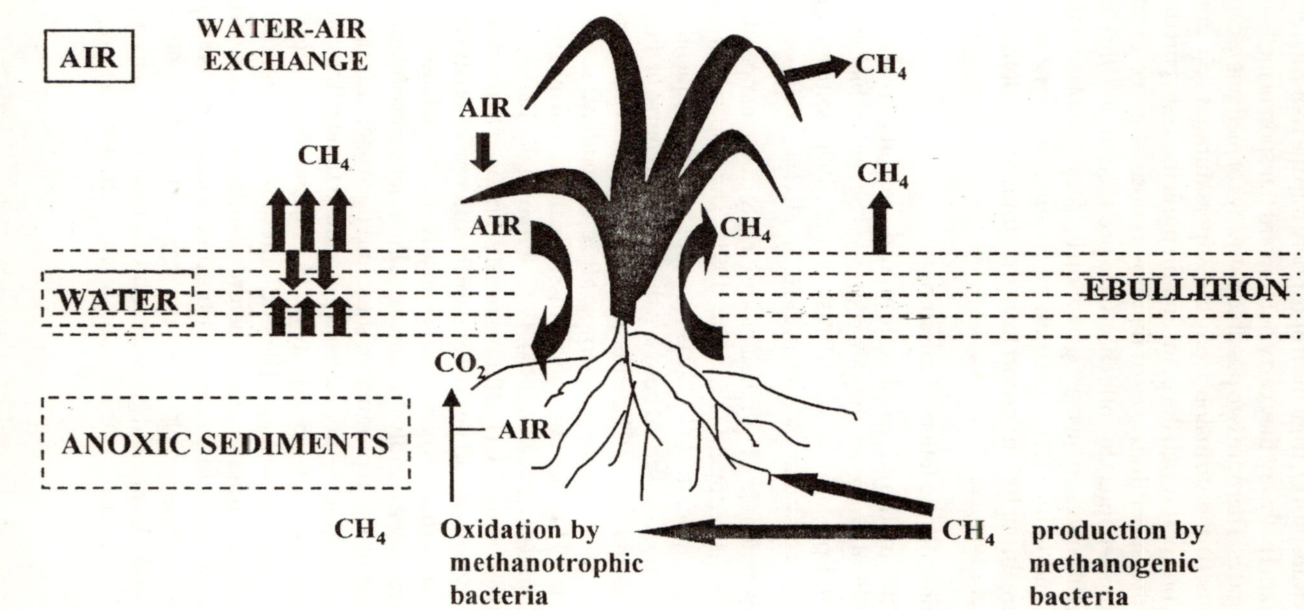

Fig. 1: Depicts the various pathways of CH_4 emission to the atmosphere

plant-mediated gas transport, vascular plants rooted in anoxic sediments transport O_2 from the troposphere into the root zone and along the same pathway, they carry out upward transport of methane and other gases. There are two possible pathways of methane escape from plants (i) the diffusion from the supersaturated soil pure solution along or through the plant and (ii) the active transport due to partial pressure differences or thermo- osmosis of gases. Plants growing in wetlands do influence CH_4 production and flux by providing substrates for methanogenesis in the form of root exudates and litter and induce CH_4 oxidation by creating oxic zone in the rhizosphere through O_2 transport from the atmosphere (Holzapfel-Pshorn *et al.*, 1985,1986).

Factors Controlling Methane Emission

As the methane production and oxidation are the microbiological processes, they are controlled by several biological, chemical and physical factors (Parashar *et al.*, 1990; Minami and Neue, 1994; Boeckx and Cleemput, 1996; Singh *et al.*, 2000).

Diel variations in CH_4 emissions are largely due to temperature variation of the sediment (Sass *et al.*, 1991). The response of CH_4 efflux to soil temperature change is very rapid and consequently no phase difference is observed between the temporal course of the soil temperature and that of CH_4 emission. Although diel variations in CH_4 emission were strongly temperature dependent, seasonal variations in CH_4 production and emission followed plant growth and development (Sass *et al.*, 1990, 1992). From negligible values at permanent inundation, CH_4 efflux generally continued to rise during the vegetative phase and peaked at panicle differentiation during a period of rapid root development, probably due to increased root exudation from the rapidly growing tips. Emissions were relatively constant during the reproductive stage and decreased during late grain filling. During the period from permanent flood to the end of the reproductive stage (65-75) days, CH_4 emission was correlated with above ground biomass. The addition of readily degradable carbon (e.g. rice, grass or straw) before planting increases early-season CH_4 emission. In the early season, production is concentrated near the base of the rice plants. As the season progresses and root system extends deeper and laterally from the base of the plant, CH_4 production in these regions increases along with root biomass.

Methanogens are pH-sensitive and grow over a relatively narrow pH range (about 6-8). A pH shift beyond this range adversely affects CH_4 generation. However, CH_4 production has been also reported in acidic environment such as peat bogs (Crawford, 1984) and in alkaline soils (Oremland, 1988). According to Wang *et al.* (1993), CH_4 emission is not strongly related to air-dried soil pH. Sometimes, acidic or alkaline soils produced more CH_4 than the soils near the neutral pH. The application of rice- straw and chemical fertilizer does have influence on soil pH, which temporarily changes the CH_4 efflux.

Flooded soils are characterized by lack of sufficient O_2 in the soil atmosphere to act as the soil electron acceptor for microbial, plant and animal respiration (Reddy and Patrick, 1984). In fact, redox potential (Eh), which measures the ability of soil environment to supply electrons to an oxidizing agent or to take up electrons from a reducing agent, determines the initiation of methanogenesis. In a soil ecosystem, Eh is determined by a number of soil chemical and biochemical reactions involving oxidation and reduction processes. As the methanogenesis occurs in the strict anaerobic environment, CH_4 generation is found to be associated with redox state of the soil. The critical Eh for CH_4 production was observed in the range of – 140 to – 160 mV (Masscheleyn *et al.*, 1993; Wang *et al.*, 1993). An exponential relationship was observed between CH_4 production and soil Eh when it was lowered below –150 mV.

Presence of mineral oxidants after the depletion of O_2 determines the redox status of the soil. Before methane production is initiated in the sediment after flooding, reductions of NO_3^- to NO_2^- and N_2O to N_2, Mn^{4+} to Mn^{2+}, Fe^{3+} to Fe^{2+}, SO_4^{2-} to S^{2-} and CO_2 to CH_4 do occur sequentially in the soil (Patrick and Delaune, 1977). A corresponding decrease in Eh indicates clearly the depletion of the subsequent oxidants. Sulphate reduction is the last step in the sequential reductions to occur at as low as –150 mV when CH_4 formation begins (Connell and Patrick, 1969; Jakobsen *et al.*, 1981).

Methane Oxidation

CH_4 emission is the net result of methane formation and oxidation. Hence, CH_4 oxidation plays a very important role in CH_4 efflux from the rice fields. About 80% of the CH_4 generated in the sediment is oxidized in different oxic layers and remaining is released to the atmosphere through vascular plants and other processes. Field experiments have shown the ratio of methane emission to production in the range of 3 to 56% in Italy (Schutz *et al.*, 1989) 10 to 81% in Texas (Sass *et al.*, 1990) and about 16% in Hunan.

Methane can be oxidized by aerobic and anaerobic bacteria. Although anaerobic oxidation is less understood, but appears to be an important CH_4 sink in certain types of anaerobic environment (Alperin and Reeburgh, 1984). A recent report indicates that methanogens have limited capacity to reverse their normal metabolism, thereby oxidizing CH_4, rather than producing it from H_2 and CO_2 (Delong, 2000). However, aerobic oxidation is well understood. Methanotrophs responsible for CH_4 consumption are seated in the oxic zones of the flooded soils, preferably in the rhizosphere and do require little oxygen for their growth (Cicerone and Oremland, 1988). Bender and Conrad (1994) reported CH_4 oxidation activity strictly coupled to the presence of O_2, indicating low importance of anaerobic oxidation. However, Czepiel *et al.* (1996) observed that CH_4 oxidation in pre-incubated paddy soils was insensitive to O_2 mixing ratios higher than 1-3%. Thus, methanotrophs have to compete with heterotrophic bacteria for available O_2. Especially in CH_4–rich environment, methanotrophs out compete other O_2 utilizing micro-organisms (Bender and Conrad, 1994).

In paddy fields, molecular O_2 is rapidly consumed after flooding. Atmospheric O_2 transport via the aerenchyma of aquatic plants to the roots is used in respiration by the plant roots. During the flooding season, O_2 is present only in the flood water-soil interface, the surface layer of paddy soils and probably in the rhizosphere of rice plants. Therefore, CH_4 oxidation can take place only in the water-soil interface and in rhizosphere during the flooding period. In fact, oxidation of CH_4 in paddy soil plays an important role in controlling the CH_4 emission from the rice fields on drainage of water.

De Bont *et al.* (1978) counted 10 times more methane oxidizing bacteria in the rhizosphere than in the bulk anaerobic soil and 113 times more than in the oxidized soil water interface. Hence, oxidation of CH_4 is expected to be very high in the rhizosphere. However, incubation experiments of CH_4 oxidation in various soils show that the potential oxidizing activity was distributed heterogeneously with soil depth. In addition to maximum oxidation at 0-1 cm in paddy soil, there are several sub maxima in the deeper layers, suggesting the presence of oxic zones in the anoxic layer (Bender and Conrad, 1994). De Bont *et al.* (1978) reported a significant increase in CH_4 emission when CH_4 oxidation was inhibited by acetylene exposure at the soil water interface. However, the effect was not obvious when acetylene was applied to the rhizosphere. Hence they suggested that CH_4 oxidation in the rhizosphere was of minor significance.

A number of factors modulate the CH_4 oxidation in the paddy soils. Dorr *et al.* (1993) established a good correlation between CH_4 oxidation and soil permeability. In fact, soil permeability is determined by several factors like soil texture, porosity, soil moisture, organic carbon content etc. A strong dependence of CH_4 consumption on moisture content was proved by the incubation experiments in temperate zones. The oxidation rate maxima was observed at different moisture contents in different soils and ranged from 18- 33% for the grassy area soil and from 30-51% for wooded area soil (Czepiel *et al.*, 1996). Uptake of CH_4 in desert soil was also enhanced by rainfall after an initial drainage (Striegel *et al.*, 1992). This underlined the importance of moisture in CH_4 oxidation, which may be linked to microbial proliferation (Whalen and Reeburg, 1990). A rapid increase in methane oxidation rates was observed over 5°C to 36°C temperature range under unlimited CH_4 supply in which the methane monooxigenase activity (MMO) of methanotrophs becomes the limiting factor (Adamsen and King, 1992; Czepiel *et al.*, 1996). However, under limited supply of methane, the temperature does not influence CH_4 uptake (Dorr *et al.*, 1993). The optimum temperature for CH_4 oxidation ranges between 31°C to 38°C for different soils. Besides, CH_4 oxidation activity has been correlated with the size of mineral particles, because of the preference of methanotrophs for the coarse-grained soils (Born *et al.*, 1990; Bender and Conrad, 1993).

The conditions of CH_4 oxidation may well explain why there is negligible CH_4 emission from upland soils. With the change from anaerobic to aerobic conditions, the CH_4 consumption is greatly enhanced on drainage of water (Keller *et al.*, 1990). However, CH_4 oxidation is suppressed by the use of ammonia based fertilizers as observed in field and laboratory conditions (Conrad and Rothfuss, 1991).

Mitigating Strategies

The paddy fields are one of the major sources of methane emission. Khalil and Rasmussen (1990) estimated a budget of CH_4 emission from rice paddies in the range of 18-120 Tg/yr, which corresponded to between 10-70% of the total anthropogenic methane. India and China, both major rice producing countries of Asia, are thought to be major contributors to global methane budget. Environmental Protection Agency, USA has projected a figure of 37.8 Tg/yr CH_4 for the Indian rice fields, which cover 42.8 m ha of arable land. However, a coordinated study under the umbrella of *Methane Campaign*, 1991 on CH_4 efflux from paddy fields in different agro-ecological regions categorically proved that the Indian paddy fields produced only 4 Tg/yr CH_4 which was only one-tenth of the figure projected by EPA. However, we have not to be complacent with this finding. In view of impending danger of climate change associated with unabated increase of mixing ratio of CH_4 in the atmosphere, there is growing need to develop certain inexpensive strategies to contain CH_4 emission from rice fields.

Water Management

Methane emission rates vary markedly with water regimes. A single mid season drainage may reduce CH_4 emission by about 50% (Kimura, 1992; Sass *et al.*, 1992). Multiple aeration for 2-3 days at 3, 6 and 9 weeks after initial flooding could mitigate CH_4 emission by 88%, without affecting rice yields as compared to the normal irrigation in southern USA (Sass *et al.*, 1992). On the contrary, Wang (1986) observed improved rice yields when short aeration is done at the end of tillering stage and just before heading. Only disadvantages associated with multiple aeration is that it requires 2.7 times more water than the normal flood water treatment. Neue *et al.* (1994) observed a spurt in CH_4 efflux when soil micropores start drying due to sudden release of CH_4 trapped in the pores. A net reduction in CH_4 emission is only achieved when soils become fully aerated. Besides, increasing percolation rates of water supply provide enough oxygen to raise redox potential, which decreases CH_4 production and stimulates CH_4 oxidation. Percolating water also carries organic solutes and dissolved gases into the subsoil and groundwater where leached CH_4 may be oxidized or released to the atmosphere (Kimura, 1992).

Paddy fields are commonly drained for 14 days before maturity of crop to facilitate easy harvesting wherever possible. However, there are constraints to implementing delayed flooding and mid-season drainage. The timing must be correct and water be available for reflooding. During monsoon, rice fields are naturally flooded in tropical Asia and fully controlled drainage is often not possible. In the dry season, irrigation, water becomes scarce and expensive. Water deficit during reproductive phase causes high sterility; while during vegetative stage it reduces plant height, tiller number, leaf area and also yields if the plants do not recover before flowering.

Organic Amendments

Methanogenesis is the last step in the electron transport chain of the fermentative decomposition of organic matter. Readily mineralizable soil organic matter is the main source to drive CH_4 formation in wetland rice. Vermoesen *et al.* (1991) observed a strong correlation between water soluble carbon and CH_4 generation. Incorporation of organic waste to flooded soils stimulates CH_4 production and emission (Schutz *et al.*, 1989; Cicerone *et al.*, 1992; Neue *et al.*, 1994), by enhancing reduction of soils and supplying C sources. Based on high C/N ratio, rice straw or green manures stimulate CH_4 production and fermented manures such as compost produce less CH_4 per unit carbon. Agnihotri *et. al.* (1999) also noted an enhancement in CH_4 emission from rice soils amended with rice straw and biofertilizer. However, with the treatment of fermented manures of cow dung and leaf, there was no change in CH_4 emission pattern as compared to non-amended rice soils (Fig. 2). Not only the quantity, but quality of the organic material also influences CH_4 production and emission. Composted material with high degree of humification only marginally increases CH_4 generation (Yagi and Minami, 1990; Minami and Neue, 1994). Incorporation of sludge from biogas plants into the paddy fields reduced CH_4 emission by 60% compared to unfermented manures (Wassman *et al.*, 1994). Recycling of crop residue in rice field through biogas plants also limits CH_4 emission and provides additional energy source. Hence, recycling and addition of organic substances rich in easily decomposable carbon has to be minimized in order to mitigate CH_4 emission from wetland rice fields, although organic matter is a good soil conditioner and sustains the soil fertility.

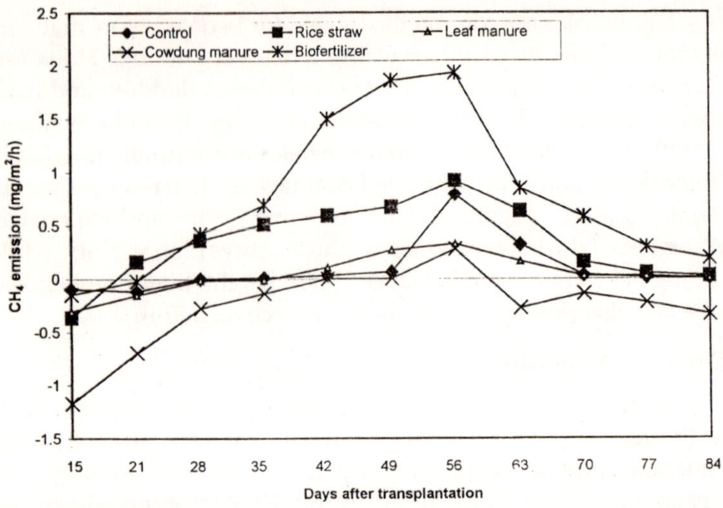

Fig. 2: Effect of organic amendments of paddy soils on CH_4 efflux

Chemical Fertilizers

Application of chemical fertilizers to agriculture fields has increased manifold to achieve the targeted food production. Effect of N-based chemical fertilizers on CH_4 production has been studied in details by Wang *et al.* (1992). Application of urea does not increase net production of CH_4, but stimulates CH_4 generation in acidic soil possibly because of a short term increase in soil pH after urea hydrolysis and a resulting decrease in Eh. Addition of ammonia to paddy fields stimulates CH_4 efflux by inhibiting CH_4 oxidation in soil-flood water interface (Conrad and Rothfuss, 1991). On the other hand, addition of NO_3–containing fertilizers decreases CH_4 production and emission by increasing the soil Eh. Similarly, sulfate-containing fertilizer competitively inhibits methanogenesis, resulting in decreased CH_4 generation. Schutz *et al.* (1989) reported a decrease of 6% CH_4 emission when $(NH_4)_2SO_4$ was added to soil surface and upto 62% when incorporated in the soil. The magnitude of decrease of CH_4 emission by SO_4–containing N fertilizer depends on the reoxidation of sulfides. Application of nitrification inhibitors like acetylene, nitrapyrin and dicyanidiamide also inhibit

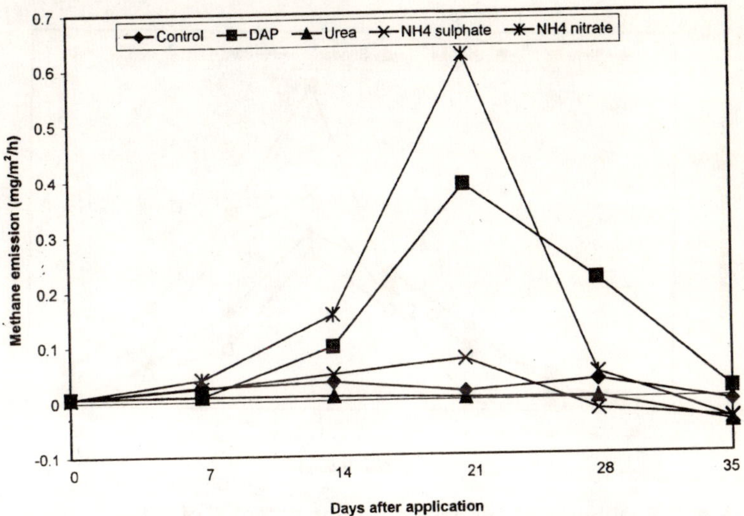

Fig. 3: Influence of application of various chemical fertilizers to paddy fields on CH_4 emission

methanogenesis and CH_4 oxidation. Slow release of acetylene from calcium carbide, encapsulated in fertilizer granules greatly reduced CH_4 emission (Bronson and Mosier, 1991).

In a field experiment with paddy crop (var. Sarju 52) fertilized with various chemical fertilizers like urea, DAP, ammonium sulphate and ammonium nitrate to paddy fields, we also observed a decrease in CH_4 emission with $(NH_4)_2SO_4$ and urea application and an increase with DAP and NH_4NO_3 (Fig. 3). It is possible that $(NH_4)_2SO_4$ competitively inhibited CH_4 production and NH_4NO_3 and DAP enhanced CH_4 emission by suppressing CH_4 oxidation. In fact, urea application suppressed methanogenesis on its hydrolysis in near neutral soils. Thus, on the basis of above studies, application of sulfate containing N fertilizers and CH_4 inhibitors may be suggested to the farmers to contain CH_4 emission from the rice fields.

Rice Cultivars

Plant-mediated CH_4 emission is a major pathway for CH_4 transport to the atmosphere. About 90% of the CH_4 released from the paddy fields is emitted through rice plants (Seiler, 1984; Holzapfel-

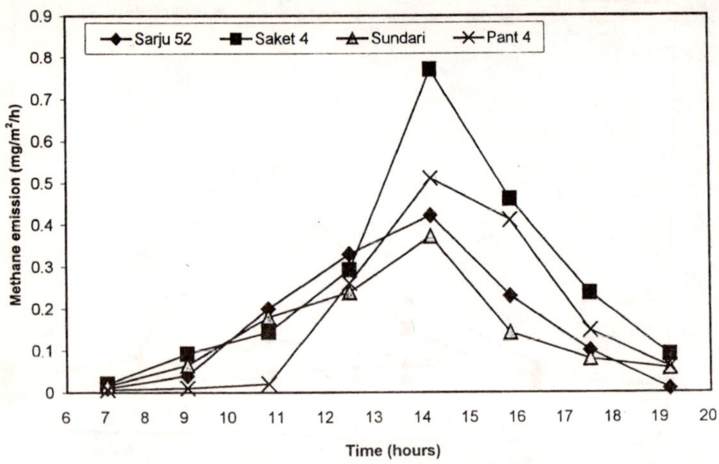

Fig. 4: Evaluation of methane emission potential of some common rice cultivars at heading stage

Pschorn *et al.*, 1986). Well developed aerenchyma in rice plants provides an efficient gas exchange between the atmosphere and the anoxic soil. In a rice plant, cell layers separating root and shoot aerenchyma determines the diffusion coefficient (Butterbach-Bahl, 1993). Cutting shoots above the water surface does not affect the CH_4 emission at all. However, number of tillers and CH_4 concentration in the medium were positively correlated with CH_4 emission rates (Mariko *et al.*, 1991). Similar relationship has been also found between root biomass and CH_4 production and with above ground biomass or grain yield (Sass *et al.*, 1990; Neue and Roger, 1993). Rice roots are the major source for CH_4 production at later growth stage because of root exudation and decay. Cultivar variation in root exudation has been also reported by Ladha *et al.* (1986).

Diffusion of oxygen from the roots facilitates the CH_4 oxidation by the methanotrophs in the rhizospere. About 90% of the CH_4 produced in the sediment is oxidized in the rhizosphere and at soil-water interface (Frenzel *et al.*, 1992). Large cultivar differences in CH_4 emission have been attributed to size of aerenchyma, root oxidation power and root exudation (Neue and Roger, 1993; Parashar *et al.*, 1990). A wide diversity of morphological and metabolical traits of rice cultivars provides an opportunity to develop high yielding and low emitting cultivars through breeding and biotechnology.

A selection of low emitting methane rice cultivars may also help in mitigating CH_4 emission from rice fields. In our study with four varieties of rice i.e. Sarju 52, Saket 4, Sundari and Pant 4 for CH_4 emission potential, Sundari was found to be lowest CH_4 emitting and Saket 4 highest emitting variety (Fig. 4). Cultivation of low CH_4 emitting rice cultivars in irrigated paddy fields will definitely reduce the methane budget.

Cultural Practices

Methane emission is enhanced through ebullition by mechanical disturbances of flooded soils in cultural practices i.e. land preparation, transplantation weeding, fertilization and harvest (Neue *et al.*, 1994). Less disturbances and shorter flooding periods in direct-seeded rice lower CH_4 emissions considerably.

Besides, crop diversification may be also one of the options to contain CH_4 emission from paddy fields (Neue *et al.*, 1991). In rice growing areas with round the year irrigation, sequential cropping of one upland crop after one or two crops may be adopted. In China, rice-wheat cropping has occupied about 10% of the arable land (Huke *et al.*, 1993). Other common crops which are widely grown in rotation with rice and mungbean, soybean, corn and vegetables. Crop intensification and diversification become possible with high investments in irrigation systems and the breeding of early maturing photoperiod-insensitive rice cultivars, which mature 60 days earlier than traditional cultivars. However, replacement of wetland rice by upland crops is hardly possible in the wet season because fields are flooded with water.

Reducing CH_4 emission from paddy fields covering a large area in different agro-ecological is no doubt an uphill task, but water management, organic amendments, fertilization, cultural practices and rice cultivars are the promising options, which may reduce atmospheric burden of CH_4 to post pone the threat of global warming.

NBRI's Contribution

We, working at National Botanical Research Institute, Lucknow - premier Institute of Botany in India, participated in "Methane Campaign, 1991 " coordinated by Dr. A.P. Mitra, Ex DG,

CSIR to measure methane efflux in different agro-ecological regions in order to compute a yearly methane budget for the Indian rice fields. This campaign was initiated in the backdrop of India being surmised to be a major contributor to global methane budget in view of 42.8 m ha of arable land under paddy cultivation. This study categorically proved that Indian paddy fields only contribute 4 Tg/yr CH_4 to global budget as against the projected figure of 37.8 Tg/yr CH_4 by EPA (USA).

Working on methane campaign enthused us to pursue further R&D work in this area. Investigating a DST–sponsored project on methane efflux from water bodies, we found seasonal dynamics in methane efflux from natural and man-made water bodies. But methane efflux was always many times higher in natural wetlands than in man-made. A marked difference in CH_4 efflux was also observed between the vegetated and non-vegetated surfaces of the same wetland, which underlined the importance of vegetation in CH_4 transport from the anaerobic soil to the atmosphere. Based on the CH_4 efflux in different seasons, we computed a yearly methane budget of 2.38×10^5 kg/yr CH_4 for local water bodies. A relationship between the CH_4 emission rates and edaphic factors was also explored in the field and laboratory conditions. Besides, CH_4 transport pathway through *Scirpus* plants was also elucidated. These findings have been already published in the journals of international repute. In 2000, a edited volume on 'Trace Gas Emissions and Plants' in incorporating recent advances in this area was also published by me from Kluwer Academic Publishers, The Netherlands. This book has been widely acclaimed by the scientists round the world.

Currently, we are working on greenhouse emission from agricultural fields and wetlands in a project sponsored by D.O.En., New Delhi. Under this project, we are measuring CH_4 and N_2O efflux simultaneously from agricultural fields and wetlands. We have also to find out the options to contain the emission of these gases. Whether N_2O transport is plant-mediated or not like CH_4 is the question to be answered. Thus we are actively involved in pursuing research in this area to come out worthwhile findings in years to come.

Conclusion

Reducing the uncertainties of regional and global CH_4 emission rates and predicting CH_4 emission trends from paddy fields requires upgraded technology and modeling of CH_4 fluxes in combination with a geographic information system of controlling factors. Rice fields are evidently a major source of CH_4 emission and so CH_4 abundance would continue to grow in the atmospheric with rise in rice production. Promising mitigating options are water management, organic amendment, fertilization, selection of suitable cultivars and cultural practices. The current understanding of factors controlling CH_4 production and oxidation is enough to develop additional mitigation technologies. But main constraints are to implementing mitigating options as the most farmers are poor and most rice growing areas need more rice to feed fast multiplying population. The farmers would adapt mitigating options only when rice production continues to increase to meet our growing demand.

References

Adamsen, A.P.S. and King, G M (1992). Methane Consumption in Temperate and Subarctic Forest Soils: Rates, vertical zonation and responses to water and nitrogen. *Appl. Microbiol.* 59: 485-490p.

Agnihotri, S., Kulshreshtha K. and Singh, S.N. (1999). Mitigation Strategy to contain Methane Emission from Rice-fields. *Environ. Monitor Assess.* 58: 95-104p.

Alperin, M.J. and Reeburgh, W.S. (1984). Geochemical observation supporting anaerobic methane oxidation, In: Microbial Growth on C-1 Compound. R.L. Crawford and R.S. Hanson (eds.), pp 282-289p, *Am. Soc. Microbiol.*, Washington, D.C.

Bekki, S., Law, K.S. and Pyle, J.A. (1994). Effect of ozone depletion on atmospheric CH_4 and CO concentrations. *Nature* (London). 371: 595-597p.

Bender, M. and Conrad, R. (1993). Kinetics of CH_4 oxidation in oxic soils. *Chemosphere* 26: 687-696p.

Bender, M. and Conrad, R. (1994). Methane oxidation activity in various soils and freshwater sediment: Occurrence, characteristics, vertical profiles and distribution on grain size fractions. *J. Geophys. Res.* 99(D8): 16531- 16540p.

Boeckx, P. and Van, Cleemput O. (1996). Methane oxidation in a neutral landfill cover soil: influence of moisture content, temperature and nitrogen- turnover. *J. Environ. Qual.* 25: 178-183p.

Born, M., Dorr, H. and Levin, I. (1990). Methane consumption in aerated soils of the temperate zone. *Tellus.* 42B: 2-8p.

Bronson, K.F. and Mosier, A.R. (1991). Effect of encapsulated calcium carbide on dinitrogen, nitrous oxide, methane and carbon dioxide emissions in flooded rice. *Biol. Fertil. Soils* 3: 116-120p.

Butterbach-Bahl, K. (1993). Mechanisms of methane production and emission in rice fields. Ph.D. Thesis, Fraunhofer Institute for Atmospheric Environmental Research, 82467, *Garmisch-Partenkirchen*, Germany (in German).

Cicerone, R.J. and Oremland, R.S. (1988). Biochemcial aspects of atmospheric methane. *Global Biogeochem Cycles* 2: 299-327p.

Cicerone, R.J., Delwiche, C.C., Typer, S.C. and Zimmermann, P.R. (1992). Methane emissions from California rice paddies with varied treatments. *Global Biogeochem Cycles* 6(3): 233-248p.

Connel, W.E. and Patrick, W.H. Jr. (1969). Reduction of sulfate to sulfide in waterlogged soil. *Soil Sci. Soc. Am. Proc.* 33: 711-715p.

Conrad, R. and Rothfuss, F. (1991). Methane oxidation in the soil surface layer of a flooded rice field and the effect of ammonium. *Biol. Fert. Soils* 12: 28- 32p.

Crawford, W. (1984). Methane production in Minnesota peat lands. *Appl. Environ. Microbiol.* 47: 1266-1271p.

Crützen, P.J. (1991). Methane's sinks and sources. *Nature.* 350: 380-381p.

Czepiel, P.M., Mosher. B., Crill, P.M. and Harris, R.C. (1996). Quantifying the effect of oxidation on landfill methane emissions. *J. Geophys. Res.* 101 : 16721-16729p.

De Bont, J.A.M., Lee, K.K. and Bouldin, D.F. (1978). Bacterial oxidation of methane in a rice paddy. *Ecol. Bull.* 26: 91-96p.

DeLong, E.F. (2000). Resolving a methane mystery. *Nature.* 407: 577-579p.

Delwiche, C.C. and Cicerone, R.J. (1993). Factors affecting methane production under rice. *Global Biogeochem. Cycles* 7: 143-155p.

Dorr, H., Katruff, L. and Levin, I. (1993). Soil texture parameterization of the methane uptake in aerated soils. *Chemosphere.* 26: 696-713p.

Frenzel, P., Rothfuss, F. and Conrad, R. (1992). Oxygen profiles and methane turnover in a flooded microcosm. *Biol. Fert. Soils* 14: 84-89p.

Holzapfel-Pschorn, A., Conrad, R. and Seiler, W. (1985). Production, oxidation and emission of methane in rice paddies. *FEMS Microbiol Ecol* 31 : 343-351p.

Holzapfel-Pschorn, A., Conrad, R. and Seiler, W. (1986). Effects of vegetation on the emission of methane from submerged paddy soil. *Plant Soil.* 92: 223-233p.

Houghton, J.T., Callander, B.A. and Varney, S.K. (eds.) (1992). *Climate change* 1992. The supplementary report to the IPCC scientific assessment. Cambridge University Press, Cambridge.

Huke, R.E., Huke, E., Woodhead, T. and Huang, J. (1993). *Rice-wheat atlas of China.* International Rice Research Institute, Los Banos, Philippines.

Jakobsen, P., Patrick, W.H.Jr. and Williams, B.G. (1981). Sulfide and methane formation in soils and sediments. *Soil Sci.* 132: 279-287p.

Keller, M., Mitre, M.E. and Stallard, R.F. (1990). Consumption of atmospheric methane in soils of Central Panama: Effect of agriculture development. Global Biogeochem. *Cycles* 4: 21-27p.

Khalil, M.A.K. and Rasmussen, R.A. (1987). Atmospheric methane: trends over the last 10,000 years. *Atmos. Environ.* 21: 2445-2452p.

Khalil, M.A.K. and Rasmussen, R.A. (1990). Atmospheric methane: recent global trends. *Env. Sci. Tech.* 24: 549-553p.

Kimura, M. (1992). Methane emission from paddy soils in Japan and Thailand. In: Batjes NH, Bridges EM (eds.) *World inventory of soil emission potentials.* WISE Rep 2, International Soil Reference and Information Centre (ISRIC), Wageningen, Netherlands. 43-79p.

King, G.M. and Adamsen, P.S. (1992). Effects of temperature on methane consumption in a forest soil and in pure cultures of the methanotroph *Methylomonas rubra. Appl. Environ. Microbiol.* 58: 2758-2763p.

Ladha, J.K., Tirol, A.C., Darby, L.G., Caldo, G., Ventura, W., and Watanabe, L. (1986). Plant associated N_2-fixation (C_2H_2-reduction) by five rice varieties and relationship with plant growth characters as affected by straw incorporation. *Soil Sci. Plant Nutr.* 32: 91-106p.

Mariko, S., Harazono, Y., Owa, N., and Nouchi, I. (1991). Methane in flooded soil, water and the emission through rice plants to the atmosphere. *Environ. Exp. Bot.* 31: 343-350p.

Masscheleyn, P.H., DeLaune, R.D. and Patrick, W.H.Jr. (1993). Methane and nitrous oxide emission from laboratory measurements of rice soil suspension: effect of soil oxidation-reduction status. *Chemosphere.* 26. 251-260p.

Minami, K. and Neue, H-U (1994). Rice paddies as a methane source. *Climate Change* 27: 13-26p.

Neue, H-U and Roger, P.A. (1993). Rice agriculture: factors controlling emission. In: Khalil, M.A.K., Shearer, M. (eds.) Global atmospheric methane: sources. sinks, and role in global change. NATO ASI Ser I, *Global Environmental Change Vol 13.* Springer, Berlin, 254-298p.

Neue, H-U, Lantin, R., Wassmann, R., Aduna, J.B., Alberto, C.R. and Andales, M.F. (1994). Methane emission from rice soils of the Philippines. In: *CH_4 and N_2O.* National Institute of Agro-Environmental Sciences, Tsukuba, Japan, 55-63p.

Neue, H-U, Quijano, C.C. and Scharpenseel, H.W. (1991). Conservation and enhancement of wetland soils. In: *Evaluation for sustainable land management in the developing world. Vol 2*: technical papers. IBSRAM Proc No 12(2), Bangkok, Thailand, 273-303p.

Nouchi, I., Mariko, S., Aoki, K. (1990). Mechanism of methane transportation from the rhizosphere to the atmosphere through rice plants. *Plant Physiol.* 94: 59-66p.

Oremland, R.S. (1988). Biogeochemistry of methanogenic bacteria. In: Zehnder, A.J.B. (od.) *Biology of anaerobic micro-organisms.* John Wiley, New York, 641-706p.

Parashar, D., Rai, J.C., Gupta, P.K. and Singh, N. (1990). Parameters affecting methane emission from paddy fields. *Indian J. Radio Space Phys.* 20: 12-17p.

Patrick, W.H.Jr. and Delaune, R.D. (1977). Chemical and biological redox systems affecting nutrient availability in the coastal wetlands. *Geosci. Man.* 18: 131-137p.

Prinn, R. (1994). The interactive atmosphere: global atmospheric-biospheric chemistry. *Ambio.* 23: 50-61p.

Rasmussen, R.A. and Khalil, M.A.K. (1984). Atmospheric methane in recent and ancient atmosphere: Concentrations, trends, and inter-hemispheric gradient. *J. Geophys. Res.* 89, 11, 599-11, 605p.

Reddy, K.R. and Patrick, W.H.Jr. (1984). Nitrogen transformations and loss in flooded soils and sediments. *CRC Crit. Rev. Environ. Control.* 13: 273-309p.

Rudolph, J. (1994). Anomalous methane. *Nature.* 368: 19-20p.

Sass, R.L., Fisher, F.M., Harcombe, P.A. and Turner, F.T. (1990). Methane production and emission in a Texas rice field. *Global Biogeochem. Cycles.* 4: 47-68p.

Sass, R.L., Fisher, F.M., Harcombe, P.A. and Turner, F.T. (1991). Mitigation of methane emission from rice fields: possible adverse effects of incorporated rice straw. *Global Biogeochem. Cycle.* 5: 275-287p.

Sass, R.L., Fisher, F.M., Wang, Y.B., Turner, F.T. and Jund, M.F. (1992). Methane emission from rice fields: the effect of flood water management. *Global Biogeochem. Cycle.* 6: 249-262p.

Schütz, H., Holzapfel-Pschorn, A., Conrad, R., Rennenberg, H. and Seiler, W. (1989). A three year continuous record on the influence of daytime, season, and fertilizer treatment on methane emission rates from an Italian rice paddy field. *J. Geophys. Res.* 94: 16405-16416p.

Seiler, W. (1984). Contribution of biological processes to the global budget of CH_4 in the atmosphere. In: Klug, M.J., Reddy, C.A. (eds.) *Current perspectives in microbial ecology.* American Society for Microbiology, Washington, pp 468-477p.

Seiler, W., Holzapfel-Pschorn, A., Conrad, R. and Scharffe, D. (1984). Emission of methane from rice paddies. *J. Atoms. Chem.* 1: 241-268p.

Singh, S.N., Kulshreshtha, K. and Agnihotri, S. (2000). Seasonal dynamics of methane emission from wetlands. *Chemosphere: Global Change Science.* 2: 39-46p.

Striegel, R.G., McConnaughey, T.A., Thorstenson, D.C., Weeks, E.P. and Woodward, J.C. (1992). Consumption of atmospheric methane by desert soils. *Nature.* 357: 145-147p.

Thomson, A.M. and Cicerone, R.J. (1986). Possible perturbations to atmospheric CO, CH_4 and OH. *J. Geophys. Res.* 9(D): 10858-10864p.

Topp, E. (1993). Effects of selected agrochemicals on methane oxidation by an organic agricultural soil. *Can J. Soil Sci.* 73: 287-291p.

Vermoesen, A., Ramon, H. and van Cleemput, O. (1991). Composition of the soil gas phase. Permanent gases and hydrocarbons. *Pedology.* 41: 119- 132p.

Wang, M.X., Shangguan, X.J., Shen, R.X., Wassmann, R. and Seiler, W. (1993). Methane production, emission and possible control measures in the rice agriculture. *Adv. Atmos. Sci.* 10(3): 307-314p.

Wang, Z. (1986). Rice-based systems in subtropical China. In: Juo, A.S.R., Lowe, J.A. (eds.) *Wetlands and rice in Subsaharan Africa.* International Institute of Tropical Agriculture (IITA), Ibadan, Nigeria. 195-206p.

Wang, Z., Delaune, R.D., Lindau, C.W. and Patrick, W.H.Jr. (1992). Methane production from anaerobic soil amended with rice straw and nitrogen fertilizers. *Fert. Res.* 33: 115-121p.

Wassmann, R., Neue H-U, Lantin, R.S., Aduna, J.B., Alberto, M.C.R., Flores, M.J.F., Tan, M.J.P., Denier Van der Gon H.A.C., Hoffmann, H., Papen, H., Rennenberg, H. and Seiler, W. (1994). Temporal patterns of methane emissions from wetland rice fields treated by different modes of N- application. *J. Geophys. Res.* 99: 16457-16462p.

Whalen, S.C. and Reeburg, W.S. (1990). Consumption of methane by tundra soils. *Nature.* 346: 160-162p.

Yagi, K. and Minami, K. (1990). Effect of organic matter application on methane emission from some Japanese rice fields. *Soil Sci. Plant Nutr.* 36: 599- 610p.

CHAPTER 17

Floristic Diversity and Biotic Pressure in Shoolpaneshwar Wildlife Sanctuary, Gujarat

Pradeepkumar, G., G. Prathapasenan and Yogesh T. Jasrai

Department of Botany, Faculty of Science
M.S. University of Baroda, Vadodara - 390002

Abstract

The floristic diversity of Shoolpaneshwar Wildlife Sanctuary has been studied in greater detail. Along with the floristic studies observations on the habitats- their present status and the major biotic factors operating in the area were also made and recorded. This paper deals with the vegetational types, floristic diversity and the impact of biotic pressure on the floristic diversity of the Sanctuary and some management suggestions to improve the system.

Introduction

Biodiversity studies are gaining immense significance and are the hot topic today. The importance of biodiversity as well as the need to protect floristically rich areas is now universally accepted. Biodiversity conservation is not only for ecological and environmental rejuvenation, but also for sustainable economic development. Being the primary producers in the ecosystem and supporting hosts for many dependant species, plants deserve particular attention. It is estimated that the disappearance of one

Photo 1: A thick forest in the core area of Shoolpaneshwar Sanctuary (Near Chopadi).

Photo 2: The cleared forest areas show the sprouting of forest trees.

Photo 3: The massive growth of *Cuscuta* on a *Ficus* tree along the roadside.

Photo 4: Clearing of forest areas -see the burnt stumps.

Photo 5: A view of Mal-samot which is totally degraded.

plant species results in the loss of 10 to 15 dependant species of insect, higher animals and even other plants (Manilal, 1997). Sanctuaries and protected areas have a major role in the biodiversity conservation because these areas can serve as natural repositories for many plant and animal species characteristic to a particular region or zone. However, the protected areas are under various threats from different sources. The major threat to such natural habitats is the biotic pressure.

The state of Gujarat is very unique in having four biogeographic zones such as Indian Desert, the semi-arid zone, the Western Ghats and the coastal region. The vast range of habitats coupled with the difference in climatic and other environmental factors in these biogeographic zones are responsible for its rich natural vegetation. The ever increasing human and cattle population, the unskilled over exploitation of natural resources and related activities have led to the disappearance of many species. Thus it is imperative to stop these activities for minimizing further deterioration and to save and protect whatever is left out. This is possible only when a first hand information about the biological resources of the protected areas, their present status and what is happening in these areas is available at one point. In this direction an attempt has been made to gather such information on Shoolpaneshwar Wildlife Sanctuary.

The area

The Shoolpaneshwar Wildlife Sanctuary (SWS) an extension of old "Dumkhal Sloth Bear Sanctuary" comprises an area of 607 sq.kms. It is located between 73° 32" and 73° 54" East longitude and 21° 45" and 21° 52" North latitude. The Sanctuary forms a connecting link between Vindhya and Satpura hill ranges. The hilly terrain with a continuum of small and high hillocks is a characteristic feature of the area. It is bounded in the north by Narmada river and the proposed Sardar Sarovar, in the west by Karjan reservoir and Tarav river. Most of its eastern boundaries slowly merge into populated areas or forests of Sagbara forest range (Fig. 1). A large number of small rivulets and streams originate from the area and after traversing the area they form tributaries of Tarav, Karjan and Narmada rivers. The climate of the area is periodical and the temperature varies between 43° C and 10° C. Average rainfall of the area is about 1000 mm. The principal geological formation of this

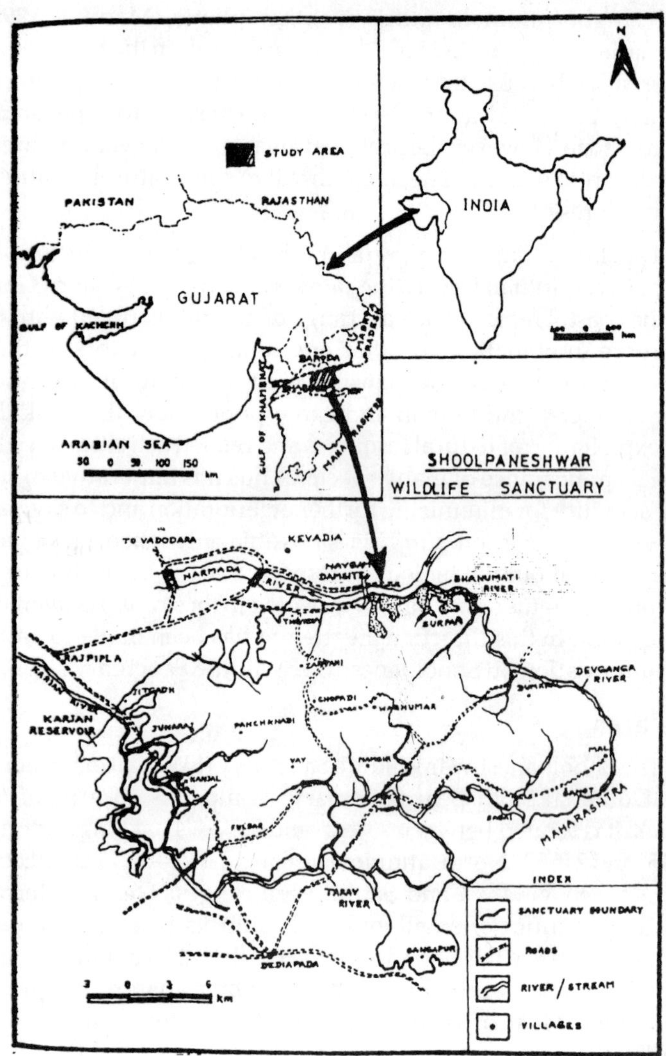

Fig. 1: Location Map

area, in sequence of deposition, are bagh limestones, sand stones, the deccan trap lava flows and the alluvial deposits. The area is highly significant because it forms a part of the catchment area of

Karjan reservoir as well as the proposed Sardar Sarovar and it also contains one of the best forests in South Gujarat.

Methodology

The area has been surveyed thoroughly by regular field trips. The field trips were arranged in such a way that the entire area was covered in all the seasons. For the convenience the entire area was divided into grids measuring 1 sq.km. each. During the survey care was taken to visit almost all the grids. The collected plant samples were properly identified, preserved following routine herbarium techniques and the data on biotic pressures were documented in the form of field notes and photographs.

Observations

The area depicts a typical dry deciduous forest, which in some parts is dominated by *Tectona grandis* and in other parts, a combination of *Tectona grandis, Haldinia cordifolia, Mitragyna parvifolia, Terminalia crenulata* and *Anogeissus latifolia*. The Northern periphery of the Sanctuary depicts a more or less thorny scrub forest dominated by *Mimosa hamata, Acacia nilotica, Euphorbia neriifolia* and *Nyctanthes arbor-tristis*. Areas such as Namgir, Kalvat, and Tukner exhibit the dominance of bamboo (*Dendrocalamus strictus*). The riverine areas support a wide variety of species viz. *Madhuca longifolia, Pongamia pinnata, Ficus racemosa, Mangifera indica, Syzygium heyneanum, Tamarix dioica, Homonia retusa, Polygonum glabrum* and *Saccharum spontaneum*. The floristic studies revealed the presence of 649 angiospermic plant species belonging to 113 families (Table 1).

Table 1 : Angiosperm diversity in Shoolpaneshwar WLS

Group	*Families*	*No. of Genera*	*No. of Species*
Dicots	91	342	510
Monocots	22	91	139
Total	113	433	649

Floristic analysis also depicted the occurrence of some rare plants (Table 2) and the dominance of a few families (Table 3).

It is imperative that Sanctuary at present embodies a good number of plant species with great diversity. However the area is under tremendous biotic pressure (Fig. 2) as evident from the fact that the area boasts 103 villages, 37,000 human population and 26,000 cattle according to the 1991 census. The tribals in the area are

Table 2 : Plants with rare distribution in SWS

Botanical Name	*Family*
Dilleni pentagyna Roxb.	Dilleniaceae
Cochlospermum religiosum (L.) Alst	Cochlospermaceae
Firmiana colorata (Roxb.) R. Br.	Streculiaceae
Schleichera oleosa (Lour) Oken	Sapindaceae
Raderemachera xylocarpa (Roxb.) K. Schum	Bignoniaceae
Phanera integrifolia (Roxb.) Bth.	Caesalpiniaceae
Butea parviflora Roxb.	Fabaceae
Colebrookea oppositifolia Sm.	Lamiaceae
Begonia crenata Dryand.	Begoniaceae
Oroxylum indicum (L.) Vent.	Bignoniaceae
Dendrobium microbulbon A. Rich	Orchidaceae
Careya arborea Roxb.	Lecythidaceae
Aeginetia indica L.	Orobanchaceae
Ensete superbum (Roxb.) Cheesman	Musaceae
Schrebera swietenioides Roxb.	Oleaceae
Sterculia urens Roxb.	Streculiaceae

Table 3 : Dominant angiosperm families in SWS

Family	*Number of species*
Poaceae	66
Fabaceae	61
Asteraceae	36
Euphorbiaceae	30
Acanthaceae	27
Cyperaceae	26
Convolvulaceae	20
Malvaceae	18
Lamiaceae	17
Amaranthaceae	17

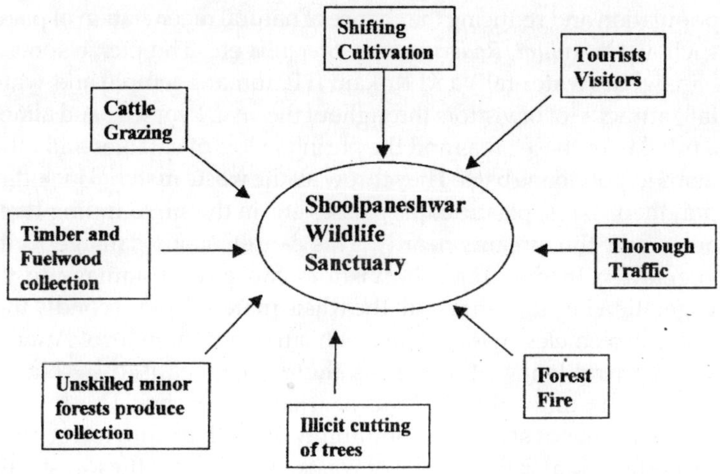

Fig. 2: The Multidirectional threats in Shoolpaneshwar Sanctuary

dependent on the forest for all their needs i.e. food, fire, furnishings for their dwelling, fodder etc. Earlier, the local populace adopted shifting cultivation i.e. they used to clear gradually the hillocks year after year. For clearing the area the trees are cut and eventually burn them (Plate 1). This is quite evident from the fact that almost all the agricultural fields in the post monsoon season shows sprouting of forest tree species such as *Butea monosperma, Terminalia cremulata* etc. (Plate 1). The cattle population in the area is another major threat to the floristic diversity. The Sanctuary has 26,000 resident cattle and in addition to this people residing in the surrounding areas of the Sanctuary bring in their cattle for grazing. This severely affect the natural regeneration of many palatable species and the intrusion of obnoxious weeds such as *Xanthium strumarium, Cassia tora* etc., into the interior forest areas and there by destroying the diversity of flora. The tribals in the area set fire to propitiate God and Goddesses. They also set fire in the month of March -April to clear the ground vegetation near *Madhuca* trees for the easy collection of flower buds. Another serious threat is the unskilled and indiscriminate method of collection of minor forest produce. They lope the twigs of trees or cut the entire plant for collecting the plant parts.

This practice lead to the complete elimination of the existing population and reducing the chance of natural regeneration of plants such as *Oroxylum, Radermachera, Sterculia* etc. The picnic spots at Ninaighat (water fall) and Kokam (Hanumanji temple and water fall) attract a lot of visitors throughout the year. People spend almost a full day in these areas and they bring in lots of eatables and other items in polythene bags. They throw all the waste material including polythene bags, plastic cups, glasses etc. in the surrounding forest areas and the streams nearby. This causes major damage to the vegetation in the area. The visitors cause dual damage to the vegetation i.e., they throw all the waste material and secondly they come in vehicles, which cause vehicular pollution also. Another serious problem in the area is the damage caused by Nature Education Camps through half-hearted approaches. This brings a large number of students' community and others into forest areas. These people also throw a lot of waste material into the forest areas and thereby damaging the system. In fact, these two components namely eco -tourism and natural camps are the two approaches for generating awareness among the masses provided these activities are well-planned and executed efficiently.

The road traffic through the Sanctuary is contributing a lot to the destruction of the floristic diversity of the area. This enhances the illegal cutting and easy smuggling of trees. It also brings in a large number of exotic plants and serious parasites. To cite an example the total stem parasite *Cuscuta* can be seen luxuriantly growing on the trees on the roadsides (plate 1). This in the later stage makes its way into the interior forest areas destroying the natural vegetation. Illegal cutting of trees and removal of natural ground vegetation enhances the rate of erosion, which in turn destroys the vegetation. The cumulative effect of all these factors has resulted in the deterioration of natural vegetation and converting the forest lands to pure agricultural fields. This is evident in Mal-samot area. The area was under good natural vegetation (Joshi 1983). The area at present is under cultivation and long stretches of agricultural fields with a few scattered trees can be seen in the area (Plate 1). If the above-mentioned factor continue to operate without any control the entire Sanctuary will gradually become pure agricultural fields as that of Mal-samot.

Conclusion

The present study was conducted to enlist the factors responsible for the deterioration of floristic diversity in Shoolpaneshwar Wildlife Sanctuary. The serious problems discussed above need to be tackled at the earliest to save the characteristic flora, fauna and natural habitats of many plants and animals. A recent visit to the area also revealed the intensity of these problems. So it is very essential to adopt some corrective measures to save the area. This area can be further protected by adopting certain corrective measures such as (1) The core area of the Sanctuary should be set free from the influence of human and cattle pressure and it should be left such. (2) Controlled grazing and stall-feeding should be encouraged. (3) Plantation in and around tribal settlements should be undertaken by incorporating the useful plants, which will reduce the load on the forests. (4) The thorough traffic should be checked. (5) The felling of trees should be checked. (6) Awareness of local people about the importance of forests, plants, and animals. These remedial measures will help to protect the area from further deterioration.

Acknowledgement

The authors wish to put on record their sincere thanks to the research team "Eco-environmental and Wildlife management studies on the Sardar Sarovar environs in Gujarat" for their help during data collection.

References

Bhatt, R.P., Bedi, S.J. and Sabnis, S.D. (1971). Botanical explorations of the Gora Range of the Rajpipla forests, Gujarat State. *Ind. For.* 98(8): 477-486p.

Cooke, T. (1901-1908). *The flora of the presidency of Bombay.* London. (Reprinted in three volumes, 1958, Calcutta).

Joshi, M.C. (1983). *A floristic and Phytochemical survey of some important South Gujarat forests with special reference to plants of medicinal and ethnobotanical interest.* Ph.D. thesis, The M.S.University of Baroda, Vadodara.

Manilal, K.S. (1997) National Parks and Conservation. A case Study of Silent Valley. In "*Conservation and economic evaluation of Biodiversity*" (Vol. 1) (Eds.) Pushpangadan, P., Ravi, K. and Santhosh, V. Oxford & IBH Publishing Co. Pvt. Ltd. New Delhi.

Pradeepkumar, G. (1993). *Vegetational and Ecological Studies of Shoolpaneshwar Wildlife Sanctuary.* Ph.D. thesis, The M.S.University of Baroda, Vadodara.

Pradeepkumar, G. and Prathapasenan, G. (1998). Studies on the Vegetation of Shoolpaneshwar Wildlife Sanctuary in Gujarat. *J. Liv. World.* 5(2): 13-21p.

Pradeepkumar, G. and Prathapasenan, G. (2000). Rare Plants of Shoolpaneshwar Wildlife Sanctuary in Gujarat. *Ecol. Env. & Cons.* 6(2): 211-215p.

Rao, R.R. (1994). *Biodiversity in India* (Floristic Aspects). Bishen Singh, Mahendrapal Singh Pub. Co., Dehra Dun.

Sabnis, S.D. and Amin, J.V. (1992). *Eco-Environmental Studies of Sardar Sarovar Environs.* The M.S. University of Baroda, Vadodara.

Shah, G.L. (1978). *Flora of Gujarat State.* The Registrar, S.P. University, Vallabh, Vidyanagar, Gujarat.

Fig. 1: A widely growing wood rotting, Basidiomycetous, polypore fungus, *Ganoderma lucidum* used for biosorption

Chapter 18
Ganoderma: A Fungus Used in Biosorption

Arun Arya and V. M. Raole
Botany Department, Faculty of Science,
The M. S. University of Baroda, Vadodara

Abstract

The most recent development in environmental biotechnology is the microbe based sorbents for the removal and recovery of strategic and precious heavy metals from industrial waste-waters. Metal removal has been tried with species of *Aspergillus, Rhizopus, Saccharomyces, Streptoverticillium, Trelletia* and *Ustilago*. Cadmium and mercury like heavy metals are removed by *Pleurotus ostreatus*. The application and potential of aphyllophore member, *Ganoderma* as fungal biosorbent has been discussed here.

Introduction

Heavy metals as wide-spread pollutants is of great environmental concern as they are non degradable and persistent (Stratton, 1987). Many of these metals are toxic and hazardous. The term *biosorption* is defined as a process in which solids of natural origin, e.g. micro organisms alive or dead, or their derivatives, are employed for sequestration of heavy metals from an aqueous environment. The transfer of metal ions from aqueous to solid phase can be due to passive, facilitated or active transport (Muraleedharan *et al.*, 1991). Microbial biomass binds with large amounts of metals and provides a cost effective solution for industrial waste-water treatment. Biosorption is possible by both living and nonliving biomass, however, bioaccumulation is mediated by only living organisms (Garnham *et al.*, 1992). Although the oligodynamic action of some metals, next particularly Cu was known since ancient civilization, the discovery of its toxic action to fungi is credited to Prevost. Till date, research in the area suggests that it is an ideal alternative for decontamination of metal containing effluents. Biosorbents are attractive since naturally occurring biomass can be effectively utilized. Besides this, biosorption offer advantages of low operating cost, minimizes the volume of the chemical or biological sludge to be disposed, being highly efficient in dilute effluents and has no nutrient requirements. These advantages have served as a viable clean-up technology for heavy metal pollution (Krattochivil and Volesky, 1998).

It is noted that binding capabilities of certain biomass are comparable with the commercially available synthetic cation exchange resins (Wase and Foster, 1997). Biosorption mainly involves cell surface complexation, ion exchange and microprecipitation (Gadd, 1990; Muraleedharan *et al.*, 1991). Different microbes have been found to vary in their affinity for different heavy metals and hence differ in their metal binding capacities. Some biomasses exhibit preference for certain heavy metals, whereas others do not show any specific binding and are broad range (Greeene and Darnall, 1990). A potant algal biosorbent Algal sorb is developed using *Chlorella vulgaris*. Immobilized biomass of *Chlorella* adsorbed Au, Cu, Fe and Zn. Another metal sorption agent (MRA) has employed *Bacillus* Biomass, Cyanobacterium- *Spirulina*. Yeast, moss- *Sphagnum* and plant *Lemna* sp. are also used as biosorbents.

Fungi Used in Biosorption

A large number of fungi have been reported to be used as biosorbents. Metal removal has been tried earlier with *Trilletia tritici, Rhizopus* sp; *Aspergillus* sp; *Streptoverticillium* and Saccharomyces (Volesky and Tsezos, 1981). Cadmium and mercury like Heavy metals are removed by *Pleurotus ostreatus*. In this age of biotechnology potential of *Ganoderma* as a fungal biosorbent is reported.

Ganoderma is a member of Aphyllophorales it is characterized by the presence of thick hard, woody basidiocarp. Stipe is short or absent. Its basidiospores are acuminate or with flat cut end, the episporium is spiny. In a survey conducted in Science faculty authors encountered a number of fruiting bodies near Botanical garden. Each basidiocarp having a red laccate surface. The sporophore measuring 15x20 mm in size is shown in Fig. 1. The fungus has been noted to have biosorption capability. Biosorption offers an economically viable technology for isolation of heavy metals (Gupta *et al.*, 2000).

References

Gadd, G.M. (1990). *Experientia*, 46: 836-840.

Greene, B. and Darnell, D.W. (1990). *In Microbial mineral recovery* (eds. Ehrlich, H.L. and Brierley, C.L.) McGraw Hill, New York. 277-301p.

Granham, G.W., Codd, G.A. and Gadd G.M. (1992). *Environ. Sc. Technol.* 26: 1764-1770p.

Gupta, R., Ahuja, P., Khan, S., Saxena, R.K. and Harapriya, M. (2000). *Curr. Sci.* 78(8): 967-973p.

Kratochivil, D. and Volesky, B. (1998). *TIBTECH.* 16: 291-300p.

Muraleedharan, T.R., Iyengar L. and Venckobachar C. (1991). *Curr. Sci.* 61: 379-385p.

Stratton, G.W. (1987). *In Review in Environmental toxicology* (ed. Hodgson, E.). Elsevier, Amsterdom, 85-94p.

Voleskey, B. and Tsezos, M. (1981). *Separation of Uranium by Biosorption.* U.S. Patent 4,320p.

Chapter 19

Response of Wheat Plants to Simulated Acid Rain

Bhoomika Singh and M. Agrawal
Department of Botany, Banaras Hindu University, Varanasi - 221 005

Abstract

Experiments were conducted in the field to investigate the influence of simulated acid rain (SAR) treatments on morphological characters, biomass accumulation and yield of *Triticum aestivum* L. (cultivars M 213 and Sonalika). Plants were treated with different levels of SAR treatments [pH 5.6 (control), [4.5 and 3.0] twice a week for 10 weeks from 15 days after germination till grain filling (80 days age). Significant reductions in total plant length, leaf area, and component biomass were observed at pH 4.0 and 3.0 of SAR in both the plant species. Yield, however reduced significantly at pH 4.0 and below in M213 and at pH 3.0 in Sonalika. The reductions in yield were also of higher magnitude in M 213 as compared to Sonalika.

The above results suggested that SAR at low pH and below have adverse effects on the morphological characters, plant biomass, and on the yield of the two cultivars of wheat. M 213 was found more susceptible to SAR treatments as compared to Sonalika.

Introduction

Acidification of rainwater is considered as one of the most serious environmental problems faced by the industrialized nations. The acid rain problem which has been considered as North

American and European problem is also spreading in the developing countries of the world (Khemani *et. al.*, 1989). Occurrence of acid rain problem in India is often ruled out due to abundance of alkaline particles in the atmosphere. But the strict rule of maintaining air quality at source has reduced the particulate emissions. However, the gases responsible for incorporating sulfate and nitrate ions into the precipitation are not reduced. Heavy acid rain problem has been predicted in Bihar, West Bengal, Orissa and south Eastern part of the country.

Acid rain has been shown to adversely affect the crop plants (Lee *et. al.*, 1981; Craker and Waldron, 1989) as well as the forest trees (Temple, 1988). Acid rain is believed to be harmful to vegetation both through direct effect on foliage and indirectly by making the soil impoverished of nutrients (Rathier and Frink, 1984). India is an agricultural country and one of its top priorities is to raise the production of food crops in order to feed its growing population. In this study, the response of two cultivars of wheat plants to simulated acid rain of different pH was assessed.

Materials and Methods

The present study was conducted at the agricultural farm of the Institute of Agricultural Sciences, Banaras Hindu University (B.H.U.), Varanasi (25°18′N latitude and 83°01′E longitude and 76.19 m above the mean sea level in the eastern Gangetic plains of India). The soil was pale sandy loam having organic carbon of 0.48 to 0.58% and pH 7.4 to 7.9. During the experiment the minimum and maximum temperature and relative humidity varied between 7.3-19.9 and 22.7-35.7 °C and 27.6-44.2 % and 56.3-87.6 %, respectively. Rainfall during the growth period of wheat was 191 mm. The two wheat cultivars M 213 and Sonalika used for the study were recommended varieties for the study area.

The experimental design was a split plot with cultivars as the whole plots and SAR treatments as the subplots (replicated thrice), randomized within the whole plots. The experiment had three treatments of SAR i.e. pH 5.6 (control), 3.0 and 4.5 applied within open top chambers of 2x2x1.2 m^3 simultaneously, using a rain simulator after 15 days of germination of plants from 9.00 to 11.00 a.m. twice a week for 9 weeks up to 85 days age. Chambers were removed after treatments. The pH of the rain solution including control was adjusted to different pH by mixing IN H_2SO_4 and IN

HNO_3 (4:1; V:v), solutions in deionised water. SAR solutions (500ml) were applied using a nozzle mounted at 1.2 m above ground in the center of the plot. Droplet size of artificial rain averaged 1.7 mm and the intensity of rain was approximately 12mm h^{-1}, which corresponds to Indian ambient condition.

The field was prepared prior to wheat sowing following standard agronomic practices. Recommended doses of N, P, and K fertilizers i.e. 120, 80 and 40 kg ha^{-1}, respectively as urea, single superphosphate and muriate of potash were applied to each plot. A half dose of N and full dose of P and K given as basal dressing and another half dose of N was given as top dressing. Similar irrigation was followed in each plot to provide similar water regime. Density was 20 plants m^{-2}.

Six monoliths of 10x0x20 cm^3 with intact roots were carefully dug out randomly from each control (pH 5.6) and SAR (pH 4.5 and 3.0) treated plots for different growth analyses at 45 and 75 days ages and then thoroughly washed. For each treatments, two plant samples were taken randomly from each replicate plots. The morphological characteristics of plants were studied in relation to total plant length, tiller number and leaf area. Leaf area was measured using a leaf area meter (Model 3100, LICOR, U.S.A.). Plant parts were separated and after oven drying at 80°C weighed separately to calculate component wise biomass. At maturity (115 days), yield was quantified as g $plant^{-1}$ on 10 plants collected from replicate plots of each treatment.

Data were subjected to two way analysis of variance (ANOVA) to examine the effects of age, treatment and their interactions. Significant differences between treatments were determined using Duncans multiple range test at the 5% level.

Results

Visible leaf injury was observed in form of leaf peeling at pH 3.0 after attaining 75 days age. The total plant length declined significantly at both the ages at SAR pH 4.5 and 3.0 with respect to the control (Tables 1 and 2). However, level of reduction was of higher magnitude at 45 days age than at 75 days. Significant decrease was also observed in case of total leaf area of SAR treated wheat plants, the magnitude of reduction was higher at 75 days plant age. At pH 3, the per cent reductions in leaf area were 17 and 33 at 45 and 75

Table 1: Effect of different SAR treatment on morphological parameters, biomass and yield of *T. aestivum* cv. M213 (Mean ± ISE).

Parameters	*M213*					
	45			*75*		
	pH 5.6	*pH 4.5*	*pH 3.0*	*pH 5.6*	*pH 4.5*	*pH 3.0*
Total plant length (cm plant^{-1})	33.0^{a} ±1.15	28.95^{bc} ±0.30	25.83^{b} ±0.50	90.42^{b} ±2.74	85.58^{b} ±0.87	78.97^{c} ±0.87
Leaf area (cm^{2} plant^{-1})	192.67^{a} ±6.45	1666.67^{bc} ±2.42	160.33^{c} ±6.04	321.00^{a} ±13.33	276.67^{b} ±5.77	216.17^{c} ±10.43
No. of Tiller plant^{-1}	3.00^{a} ±0.00	3.17^{a} ±0.17	3.17^{a} ±0.31	3.33^{a} ±0.21	3.00^{a} ±0.00	3.33^{a} ±0.21
Root biomass (g plant^{-1})	0.19^{a} ±0.02	0.16^{ab} ±0.02	0.12^{b} ±0.02	0.66^{a} ±0.04	0.46^{b} ±0.02	0.39^{b} ±0.03
Shoot biomass (g plant^{-1})	0.58^{a} ±0.02	0.41^{b} ±0.03	0.38^{b} ±0.02	4.14^{a} ±0.25	3.19^{b} ±0.08	2.84^{b} ±0.10
Leaf biomass (g plant^{-1})	0.57^{a} ±0.02	0.46^{b} ±0.02	0.43^{b} ±0.03	1.04^{a} ±0.04	0.60^{b} ±0.02	0.50^{c} ±0.03
Yield (g plant^{-1}) at 115 days	–	–	–	5.10 ±0.14	4.33 ±0.10	3.90^{c} ±0.10

Table 2: Effect of different SAR treatment on morphological parameters, biomass and yield of *T. aestivum* cv. Sonalika (Mean ± ISE).

Parameters	*Sonalika*					
	45			*75*		
	pH 5.6	*pH 4.5*	*pH 3.0*	*pH 5.6*	*pH 4.5*	*pH 3.0*
Total plant length (cm plant^{-1})	28.83^{a} ±0.64	27.18^{ab} ±0.51	25.67^{bc} ±0.46	89.90^{a} ±1.29	85.58^{b} ±1.57	80.17^{c} ±1.70
Leaf area (cm^{2} plant^{-1})	183.67^{a} ±4.74	172.17^{b} ±6.13	155.83^{b} ±5.47	305.33^{a} ±4.54	254.33^{b} ±12.77	205.67^{c} ±3.83
No. of Tillers plant^{-1}	3.00^{a} ±0.00	3.00^{a} ±0.00	3.00^{a} ±0.00	3.50^{a} ±0.22	3.00^{a} ±0.26	3.00^{a} ±0.00
Root biomass (g plant^{-1})	0.16^{a} ±0.01	0.13^{ab} ±0.08	0.11^{b} ±0.01	0.65^{a} ±0.02	0.52^{b} ±0.03	0.38^{c} ±0.03
Shoot biomass (g plant^{-1})	0.54^{a} ±0.02	0.47^{a} ±0.03	0.33^{b} ±0.01	4.10^{a} ±0.30	3.47^{a} ±0.18	2.41^{b} ±0.17
Leaf biomass (g plant^{-1})	0.58^{a} ±0.02	0.54^{a} ±0.03	0.42^{b} ±0.03	1.01^{a} ±0.06	0.87^{ab} ±0.05	0.63^{c} ±0.03
Yield (g plant^{-1}) at 115 days	–	–	–	4.50 ±0.14	4.30 ±0.10	3.90 ±0.10

For each parameter, values not followed by the same letter are significantly different at $p<0.05$.

days ages, respectively in M213. No significant change was observed in number of tillers of SAR treated wheat plants. Component wise biomass accumulation was also adversely affected by increasing acidity of rain (Tables 1 and 2). Significant reductions were observed in shoot and leaf biomass of plants treated at pH 4.5 and 3.0 at both the ages. In case of root biomass, reductions were significant at pH 4.5 and 3.0 at 75 days and only at pH 3 at 45 days age in M213. Significant yield reductions were observed at SAR pH 4.5 and 3 in M213 and at pH 3 only in Sonalika as compared to the control (pH 5.6) (Tables 1 and 2).

Discussion

The results showed that twice-weekly application of simulated rain of pH 4.5 did not cause any foliar injury symptoms. Only SAR of pH 3 caused epidermal leaf peeling at 75 days age. Author (1984) reported visible injury below pH 3. In case of twenty soybean cultivars, visible injury occurred early in the growing season when the plant was at first and second trifoliate leaf stage (Banwart *et al.*, 1990), but in the present study symptoms appear quite late during grain filling stage of wheat. Johnston and Shriner (1985) observed leaf tip necrosis in wheat from pH 4.5 to 2.3.

The SAR at low pH adversely affected total plant length and leaf area. Johnston and Shriner (1985) reported growth reductions in wheat cultivars Abe and Arthur 71 at pH 4.0±0.5, whereas, wheat cultivar Oasis showed no growth reductions. Singh and Agrawal (1996) have also observed reduction in plant height at SAR pH of 4.5, 4 and 3 at 75 days age in *T. aestivum* var M 206. Plant height of *Zea mays*, applied biweekly with simulated rain treatments of pH 4.6 to 3.0 showed no significant effect during early stage, but at latter stage variety pioneer 3377 showed a quadratic effect with increasing acidity (Smith *et al.*, 1990). The plants at pH 5.6 were taller than the average of other treatments. However, no effect of acidity on plant height was observed for other variety P73 x M 017. Simulated acid rain significantly reduced the leaf area at both the ages. Reductions in leaf area of *T. aestivum* var M 206 and M 234 have also been reported at pH 4.5 to 3.0 (Singh and Agrawal 1996). Evans *et al.* (1982) have shown significant reductions in the leaf area of radish plants exposed at pH 4.2, 3.4, 3 and 2.6. Kohno and Kobayashi (1989) also reported reduction in leaf area of trifoliate leaves of soybean plants. Johnston *et al.*, (1982) have also reported significant reductions in leaf area of bush bean treated as pH 4 on a six day cycle consisting of two consecutive days with rain followed by four days without rain and the rain acidity was adjusted with sulphuric acid. The reduction in leaf area may be due to altered leaf development in response to acid rain due to lowering of photosynthetic capacity. Evans and Lewin (1981) have shown that increase in acidity decreases the leaf expansion both in pinto bean and soybean.

Tiller number did not show significant variation in response to increasing acidity of SAR. Johnston and Shriner (1985) showed that tillering increased with rain acidity in wheat cultivar Abe, but did not change in case of Arthur and Oasis. Pell and Puente (1987) did not find any significant effect of acidity of rain on number of tillers per hectare of *Avena sativa* L. ev. Ogle plants treated at 3, 3.6, 4.2 and 4.8 compared to no rain treatment. Biomass accumulation in different plant parts showed significant decline especially at lower pH of SAR. Evans *et al.* (1982) observed 66 and 73% decline in root fresh mass of radish grown under green house condition and treated with SAR pH 3.0 and 2.6 respectively. Johnston *et al.* (1982) also found significant decline in root weight of bush bean plants exposed to a series of solutions at pH 2.8, 3.1, 3.4, 3.6 and 4 applied for 8 minutes. Shoot weight was reported to decline significantly in *V. faba, P. multiflorus* and *P. sativum* exposed to SAR pH 3.5 and 2.5 (Ashenden and Bell, 1989). Radish plants also showed reductions in shoot biomass at pH 3.0 and 2.6 compared to pH 5.7 (Evans *et al.*, 1982). However, Smith *et al.* (1991) have found no significant impact of rain acidity between pH 3 and 5.6 in relation to shoot biomass of *G. max* cultivars Amsoy 71 and Williams 82. Singh and Agrawal (1996) found significant reductions in both above ground and below ground biomass of SAR treated wheat plants at 75 days age. The loss in photosynthetic capacity has a direct effect on foliar growth and biomass accumulation. Leaf area is found to be significantly positively correlated with leaf dry weight.

Significant yield reductions were observed in plants treated with pH 4.5 and 3.0 as compared to the control in M213 whereas yield reductions were significant at pH 3 only in Sonalika. Artificial rain of pH 4 has been shown to reduce the marketable yield of various species compared with pH 5.6 e.g. *Daucus carota, Brassica japonica,* etc (Lee *et. al.*, 1981), however, a further decrease in pH does not necessarily result in further yield reductions. Evans *et. al.*, (1982) did not find any significant effect on yield of wheat plants exposed to simulated rain at pH between 5.7 and 2.7 from anthesis to maturity, whereas for tropical wheat cultivars M 234 and M 206, yield reductions were observed at pH 5, 4.5, 4 and 3 (Singh and Agrawal, 1996).

The study clearly indicates that simulated acid rain has a negative impact on morphological characters of wheat plants leading to biomass and yield reductions. Yield reductions showed linear response to the increasing acidity of the rainfall. Cultivar M213 showed more sensitivity to low acidity of SAR than Sonalika.

Acknowledgement

The authors thankfully acknowledge the financial assistance received from Ministry of Environment and Forest, Government of India, New Delhi.

References

Amthor, J.S. (1984). Does acid rain directly influence plant growth? Some comments and. observations. *Env Poll.* 36: 1-6p.

Ashenden, T.W. and Bell, S.A. (1989). Growth responses of three legume species exposed to simulated acid rain. *Env Poll.* 62: 21-29p.

Banwart, W.L., R.L. Finke, P.M., Porter and J.J., Hassett. (1988). Sensitivity of twenty soybean cultivars to simulated acid rain. *J. Environmental Quality.* 19: 339-346p.

Craker, L.E. and Waldron, P.F. (1989). Acid rain and seed yield. reduction in corn. *Journal of Environmental Quality.* 18: 127-129p.

Evans, L.S., Gmur, N.F. and Mancini, D. (1982). Effects of simulated acid rain on yields of *Raphanus sativus, Lactuca sativa, Triticum aestivum* and *Medicago sativa. Envi Exp Bot.* 22: 445-453p.

Evans, L.S. and Lewin, K.F. (1981). Growth development and yield response of pinto beans and soybeans to hydrogen ion concentrations of simulated acidic rain. *Env Exp Bot.* 21: 103-113p.

Johnston, J.W.Jr. and Shriner, D.S. (1985). Responses of three wheat cultivars to simulated acid rain. *Environmental and Experimental Botany.* 25: 349-353p.

Johnston, J.W.Jr., Shriner, D.S., Klarer, C.J. and Lodge, D.M. (1982). Effect of rain pH on senescence, growth and yield of bush bean. *Env Exp Bot.* 22: 329-337p.

Khemani, L.T., G.A. Momin, P.S.P. Rao, P.D. Safai, G. Singh, and R.K. Kapoor (1989). Spread of acid rain over India. *Atmos. Environ.* 23: 757-762p.

Kohno, Y. and Kobayashi, T. (1989). Effect of simulated acid rain on the growth of soybean. *Water, Air and Soil Pollution.* 43: 11-19p.

Lee, J.J., Neely, G.E., Perrigan, S.C., and Growthaus, L.C. (1981). Effects of simulated sulphuric acid rain on yield, growth and foliar injury of several crops. *Env Exp Bot.* 21:171-185p.

Pell, E.J. and Punete, M. (1987). Impact of simulated acid rain on yield of a field grown oat crop. *Env Exp Bot.* 27: 403-407p.

Rathier, T.M. and Frink, C.R. (1984). Simulated acid rain: Effects of leaf quality and yield of broadleaf tobacco. *Water, Air, Soil Pollut.* 22: 389-394p.

Singh, A. and Agrawal, M. (1996). Response of two cultivars of *Triticum aestivum* L. to simulated acid rain. *Env Poll.* 91; 161-167p.

Smith, C.R., Vasilas, B.L., Banwart, W.L., Peters, D.B. and Walker, W.M. (1990). Lack of physiological response of two corn hybrids to simulated acid rain. *Environ. Exp. Bot.* 30: 435-442p.

Smith, C.R., Vasilas, B.L., Banwart, W.L., and Walker, W.M. (1991). Physiological response of two soybean cultivars to simulated acid rain. *New Phytol.* 119: 53-60p.

Temple, P.J. (1988). Injury and growth of Jeffrey pine and Giant sequoia in response to ozone and acidic mist. *Envion. Exp. Bot.* 28: 323-333p.

Section II

Pollution: Monitoring, Assessment and Solutions

CHAPTER 20

Air Pollution: Development at What Cost?

P.S. Rao
Executive Director, Gujarat Refinery, Vadodara

At the outset, I express my sincere thanks to the Indian Association for Air Pollution Control for giving me the opportunity to address this august gathering. It is really praiseworthy to organize a seminar on such a burning issue where whole of the society is involved.

Rapid industrialization has improved the standard of living but threatened the very existence of mankind. Man is going to conquer the universe with the help of technological developments but the unsustainable development has resulted in grave danger to the very existence of man.

The destruction of the earth's life support systems beyond a certain point would become irreversible. In the past, specific societies have simply vanished and destroyed themselves as a result of neglect of their natural resources. We have perhaps not learned the lessons of history, and are today creating a global crisis from which there would be no escape.

India is making extremely swift progress in terms of economic development to meet the requirement of her people. This is helping to free many people from poverty and to elevate the living standard but it has also caused pollution to reach a global scale. Now it is high time to pay due attention for abatement of pollution. Industrial progress is must for civilization but it is to be kept in mind that human race should not be in danger at the price of it. Every industry has to play its role so that their activities should not deteriorate the environment. Their profit may be marginally down but they can not deny their social collective responsibility to mankind. Otherwise this may lead to any catastrophe which was evidenced In Japan.

Gujarat Refinery – where growth is the essence of life – is the largest refinery of Indian Oil Corporation Limited. It was commissioned in 1965 with a meagre crude refining capacity of 2.0 Million Metric Tonnes per annum (MMTPA). Subsequently, the refinery has been, expanded, revamped and modernized from time to time and today it has crude refining capacity of 12.5 MMTPA.

In this march towards growth and excellence, Refinery has not forgotten its duty to preserve the environment. It has always recognized the vital importance of peaceful co- existence of industrial development with the environment.

Compared to water pollution, air pollution follows an entirely different behaviour. Once occurs, it is difficult to control. Technologies to treat air pollutants are expensive. Hence, at Gujarat Refinery, primary thrust has been given to control air pollution at source. Fuels from indigenous crude having low sulfur level are used in unit furnaces and boilers. Energy conservation measures like replacement of low efficiency furnaces, optimisation of heat-exchanger trains etc. has resulted in reduction of fuel consumption. This has helped in containing stack emissions in spite of increase in refining capacity.

Beside these measures, necessary operational facilities like tall stacks for better dispersion of emissions, CO-Boiler in FCC unit, floating roof tanks for minimization of the emission of volatile organic compounds and cyclone separators to recover FCC catalyst fines were provided in design stage itself. In addition, ambient air quality in the refinery premises is also monitored by continuous air monitoring stations.

Among the various sources contributing to air pollution, the automobile industry has emerged as the largest source of urban air pollution in the developing countries. Govt. of India has asked oil refineries to produce and supply diesel containing 'S' content to the level of 0.25% in order to reduce vehicular emission. In view of the above, nine oil refineries have set up Diesel Hydro Desulphurisation (DHDS) Unit to produce low sulfur diesel. Among them Gujarat

Refinery was the first to commission the DHDS unit in June and started producing the required quality of diesel since Aug.'99. In addition, Gujarat Refinery is also producing 100,000 MT diesel of 0.05% sulfur content for the NCR from Sept.'99. This will drastically reduce the SO_2 emission from vehicles.

Lead has well documented negative health effects, recognized by the health community. The impact of lead exposure is thought to be particularly significant in foetuses and children because of lead's low-level effects on neurological development and intelligence scores. Tetra Ethyl Lead (TEL) is added to the petrol to boost up its octane number. Govt. of India has asked to produce petrol containing very low level of lead. However, Gujarat Refinery is one step ahead and started producing entire quantity of petrol free of lead since Sept.'99.

Plants are good indicators of air pollution, and some of them do well to fix air pollutants also. In this context, increasing vegetation in the form of greenbelts is one of the potential alternatives to mitigate air pollution as plants produce oxygen, serve as a sink for pollutants. Thus, maintaining a balanced ecology has been one of our top priorities and major thrust areas. Gujarat Refinery has taken up tree plantation on top priority and about 2 lacs grown-up trees and innumerable flowering bushes are a part of its plant and township. Two green belts with Ecological Park have been developed in the up wind & down wind directions of the refinery as a pollution abatement measure.

Our efforts in industrialization/development in the next millennium should go hand-in-hand with due concern for environment protection with concrete steps in this direction. This is the only way for realization of industrial progress in the true sense. This has always been Gujarat Refinery's endeavour as exemplified by sustainable development for preservation of environment.

Now this is the opportune moment for us to assess and evaluate what we are doing to nature's bounty that we inherited from our forefathers and how we are actually demolishing the wisdom and tradition of earlier generations who built a way of life. High profits at the expense of people's health and degradation of air, water and soil of the country is anti-national. We need industrialists who not only want the country to develop, create employment, alleviate people from poverty but also who are ready to take responsibility to preserve the ambient air, water and soil quality so around their industry.

At the end, I wish the seminar a grand success and hope, through this seminar efforts of Indian Association for the Air Pollution Control will certainly imprint on your mind. I am sure when you return to your organization, you will leave no stone unturned for prevention and control of air pollution.

Chapter 21

Environmental Legislations, Policies and Governance in India

R. C. Trivedi
Advisor,
Centre for Environment Education, Ahmedabad

Introduction

The population of India will touch the figure of 1000 millions with the beginning of twenty first century and is likely to become 1650 millions by' 2050 with even effective and efficient family welfare programmes. Thus the population will exhort its pressure in all directions including the core components such as water, air, land, food, housing and other socio-economic factors. It has been therefore observed and adequately reported that the quality of environment in India has been deteriorating gradually over the past few decades as a result of rapid industrialization and simultaneous urbanization.

The developed countries faced the similar situation in the decade subsequent to the end of 2nd World War in 1945. The most of environmental degradation issues were due to indiscriminate disposal of solid and liquid waste originating from industries. The problem of air pollution from industries and automobiles was equally serious. It was therefore realized by the most of developed countries in fifties that there was need for comprehensive legislation for control and prevention of pollution. In fact the environmental episodes – London Smog and Minamata – Mercury Poisoning in Japan in 1952 — excellerated the process for early implementation and enforcement of legislative measures.

In the two decades following the above environmental episodes, the developed countries went through the process of creating adequate infrastructure for enforcement, technological development both in the fields of manufacturing processes and pollution control systems, financial linkages, public participation etc. These countries within 20 years acquired adequate information, experience and knowledge in the upcoming field of environment protection and public health. The United Nations therefore decided to have under its banner on international conference on "Human Environment" in 1972 at Stockholm. The most of the member countries were represented by the respective governmental head or by a senior person. India was represented by then the Prime Minister Mrs. Indira Gandhi.

The conference shared the experience and knowledge gained by the developed world and came out with several recommendations to tackle the environmental problem – existing and incoming. India is one of the original signatories to the solemn declaration of the conference. This declaration provided policy guidelines particularly to developing countries to act on the essential and imperative need to protect environment.

India accordingly initiated the steps in line with the content of declaration. The issue of protection of environment and conservation of natural resources thus received due attention in planning process in seventies. In fact the fourth five year plan (1968-73) and Fifth Five Year Plan (1973-78) gave explicit recognition for integrating environmental dimension into the planning and development processes.

Legislations

The Government of India subsequent to 1972 U.N. Conference initiated the steps to control environmental pollution problems.

India is the first country in the world which has provided constitutional safeguard for the protection and preservation of the environment. The Government of India in 1976 through 42nd amendment to the constitution has incorporated :

The Directive Principles of State Policy – "The state shall endeavour to protect and improve the environment and to safeguard forests and wildlife of the country."

The Article 51–A(g) states:

"It shall be the duty of every citizen of India to protect and improve the natura1 environment including forests, lakes , rivers and wildlife and to have compassion for living creatures."

The constitutional provisions are implemented through laws of the country. India has a large number of laws and regulations related to environment protection. There are more than 200 statues having bearing on environmental matters in India. However, the major legal provisions and regulations prevailing in India are:

a) The Wildlife (Protection) Act–1972 amended in 1983, 1986 and 1996.

b) The Water {Prevention and Control of Pollution) Act–1974 amended in 1988.

c) The Water Cess Act–1977, amended in 1991.

d) The Forest (Conservation) Act–1980, amended in 1988

e) The Air {Prevention and Control of Pollution) – 1981, amended in 1987.

f) The Environment Protection Act–1986

g) The Motor Vehicles Act–1938, amended in 1988.

h) The Factory Act–1948 amended in 1988.

i) The Public Liability Insurance Act–1991

j) The National Environmental Tribunal Act –1995

k) The Merchant Shipping Act –1958

l) The Maritime Zones Act –1976

m) The Coast Guard Act –1976

The Government of India has come out with several rules, regulation and notification under the specific act. Some of them are:

a) The Hazardous Waste (Management and Handling) Rules–1989

b) The Manufacture, Use, Import, Export and Storage of hazardous micro-organisms or Cells Rules–1989

c) Coastal Regulation Zone Notification–1991

d) Notification on Environmental Statement–1993

e) Notification an Environmental Clearance–1994

There are many other provisions. However, two of them are:

a) Articles 32 and 226 under Indian Common Law provide filing a write petition to supreme court and high courts of states respectively as public interest litigation.

b) Article 133 under Indian Penal Code related to causing of Nuisance and can be revoked for pollution control also.

Impact of Legal Framework

The responsibility for implementation and enforcement of the provisions of the various laws lie with the different agencies. The laws related to in inland water and air are imp1emented by Central Pollution Control Board and State Pollution Control Boards, the Forest and Wildlife Conservation Acts, through Central and State Departments of Forests, the Motor Vehicles Act by the respective state department of transport and the Factory Act by the respective state department of factory. Marine Pollution Control is under the Central Ministry of Surface Transport. The coordination among the various enforcement agencies is one of the drawbacks of the present system.

A full fledge Ministry of Environment and Forests at Central has been created in 1985 to oversee the functions related to environmental protection. It has however, limited functions. Even the separate department on environment has been established at state level subsequently. Even then all functions related to environment protection could not be coordinated. India, therefore, needs a single central authority something of the nature of Environment Protection Agency (EPA) in USA to have proper effective implementation of the existing legal structure.

The environment protection laws in India in their contents are comparable with those in any developed country. In fact the acts like water act, air act, environment protection act etc. are criminal in nature. However, the results achieved so far with most of the existing acts and rules and regulation thereof have been not encouraging. The water and air quality has been deteriorating particularly in industrial and metropolitan areas in India. There are reports reflecting this trend all over the country. The continual degradation of environment despite several rules and regulations can be attributed to several reasons of which a few are enumerated below:

a) Political will
b) Social attitude
c) Public education, awareness and participation
d) Technical constraints
e) Financial resource
f) Management skill

The most of Indian Environment Protection Acts suffer from more than one aspect indicated above. The present policies of Command and Control for pollution control have not produced the desired results. The developed countries are gradually moving away from such policies and are working on alternative routes having willingness to participate in environment protection programmes at every level. The government of India has recognized that problem of pollution has become more acute over the period despite the regulations. It has been therefore realized that the existing regulatory instruments should be combined with market based, promotional and proactive instruments.

Pollution Abatement Policy

The United Nations constituted a committee in 1983 under the chairmanship of Md. Bruntland, then the Prime Minister of Netherland to prepare a report or document on Sustainable Development. The report has come out in the form of a book titled as "Our Common Future". This book has been well received all over the world. Subsequently the above committee in 1987 was changed to World Commission on Environment and Development through U.N. resolution. It defines the sustainable development as "the development that meets the needs of the present without compromising the ability of future generations to meet their own needs."

The 2nd U.N. conference on "Environment and Development" popularly known as The Earth Summit held in 1992 at Rio-de-Janerio, Brazil had detailed deliberation on the concept of "Sustainable Development." The Agenda 21, the official declaration gives the framework for mobilization of this principle at the world level.

The Environmental policy in India for industry until now has focussed mainly on pollution control through end of pipe treatment.

The existing policy therefore ends up in wasteful use of resources. Besides industry needs additional resources to solve the pollution problem. With due recognition to the future resources and energy scenario, India as one of the participants at the Summit and also the signatory to the Document has recently tried to formulate policies for promoting sustainable development.

It is against this background that Central Ministry of Environment and Forest has issued a comprehensive policy statement for abatement of pollution, policy statement for Environment and Development and the National Conservation Strategy. The Ministry has also prepared Environment Action Programme and every state also has been directed to have the similar exercise. The policy statement has made a welcome attempt to shift the emphasis from "command and control" to practical aspects of actual implementation.

The policy statement for abatement of pollution states the integration of environmental considerations into decision making at all levels. The policy aims at:

a) Prevention of pollution at source
b) Encourage, develop and apply the best available practical technical solutions
c) Ensure that the polluter pays for the pollution and control arrangements
d) Focus on the protection of the heavily polluted areas and river stretches.
e) Involving public participation in decision making process.

The policy statement on environment and development alongwith National Conservation Strategy has set out the priorities for:

a) Conservation of natural resources–land and water
b) Prevention and control of air pollution including noise pollution
c) Industrial Development by using a mix of promotional and regulatory steps.

The Environment Action Programme covers:

a) Control of industrial pollution with emphasis on reduction at source.

b) Waste minimization and cleaner production practice.

c) Management of hazardous and toxic waste

d) Training, education, awareness, public participation and environmental management.

The Government in line with the policy has introduced a number of schemes motivating industries to take steps for control and prevention of pollution. Some of the schemes are:

a) Water Cess

b) Charges for liquid, solid and gas emissions.

c) Credit and loan at subsidized rate of interest

d) Tax rebates

e) Ecolabelling

f) Incentive to S.S.I. sector for adoption of cleaner production and technology.

Thus the adoption of policies by the industries in conjunction with compliances of legal requirement can result in achieving the clean and healthy environment .The Environment Management Plan both at the entrepreneur and government levels has to succeed for having sustainable development.

Governance

India has a very poor record on operational and management aspects. It has very good laws, rules and regulations. It has good plans and programmes. However, where India miserable fails is in its operation and management.

As the country is preparing to enter the twenty first century with tremendous tasks of recovering from the environmental and ecological constraints, it has to work jointly in cooperation and coordination with carrying capacities. We have arrived at the end of millennium with heavy backlog on pollution prevention and limited financial resources.

The situation therefore calls for active and honest participation from each player in the unending race for environment protection, health and safety and sustainable economical growth.

The governance that too efficient and effective is very vital for realizing the goals and objectives of having clean, green and health environment.

CHAPTER 22

ISO 14001: Environmental Standards

Vinod S. Patel
President, IAAPC, Baroda

Introduction

ISO 14001 is developed by International Organization for Standardization (ISO) best known for producing ISO 9000 series. ISO is Swiss-based, world-wide organization of national standards' bodies supported by over 118 member countries. ISO is not a government entity, but a private organization whose mission is developing and publishing voluntary standards for a wide range of purposes.

The new International Voluntary Standard for Environmental Management System (EMS) is now known as ISO 14001. The EMS is set up by International Organization for the public and private sectors. These EMS have a significant impact on trades in the same way as that of the ISO 9000 series Quality Management Standards.

The overall aim of this standard is to support environmental protection and prevention of pollution in balance with socio-economic needs.

It will help an organization to enhance the efficiency or its processes and thus improve its performance. The adoption of standards will also strengthen a company's environmental credibility with customers, governments and communities.

The ISO 14001 does not establish absolute requirements for environmental performance. However, compliance with applicable legislation and regulations is essential. In addition, it must commit to continual improvement of its environmental performance. These should be clearly stated in the organization's environmental policy.

What is ISO 14001? How it Works?

ISO 14001 helps an organization to establish a disciplined system for achieving stated environmental objectives that adhere to relevant legislative and regulatory requirements, reforms according to its policies and procedures, and environmental audit, ensuring full conformance and continual improvement.

In other world ISO 14001 is the Environmental Management System with specifications for use that was published in October 1996. ISO 14001 provides that frame work for developing a EMS using a standardization approach that has international acceptance, but allows for tailoring to the organization's specific need. It is a guidelines for achieving continual improvement with impetus provided by the organization itself rather that external factors.

Why was ISO 14001 Developed?

ISO 14001 provides standardization for developing and implementing an EMS that will identify, address and improve how an organization addresses its environmental impact. The standards provide a frame-work for managing an organization's environmental aspects and monitoring the degree of environmental improvement.

How an Organization Benefits from ISO 14001?

Gaining competitive advantage through meeting customer environmental demands, public relations benefits, and innovation of products/market and cost saving from action in area such as energy efficiency and waste minimization.

1. Meeting stakeholder, including insurers, bankers, regulator, employees and the local community environmental requirements.
2. Satisfying investor criteria and improving access to capital.
3. Obtaining insurance at reasonable cost.
4. Improving environmental performance.
5. Improving management assurance.
6. Improving cost and quality control.

7. Enhancing images and market share.
8. Improving industrial performance and government relation.
9. Conservation energy and input materials.
10. Meeting vendors for quality improvement and certification criteria.
11. Minimize costs.
12. A step ahead in the competion.

Is ISO 14001 Related to ISO 9000?

Both ISO 14001 and ISO 9000 have similar framework for documentation and implementation. The ISO 9000 Quality Management standards focus on systems to improve environmental performance. Organization may decide to implement either standards or both, but coordinated implementation may be a viable option for such organization.

CHAPTER 23

Environmental Performance Evaluation: ISO 14031

Yogesh. T. Jasrai
General Secretary, IAAPC, Baroda

EPE (Environmental Performance Evaluation) is the ongoing evaluation of the environmental performance of an organization. It is a method to measure the results of the organization's management of the environmental aspects of its activities, products and/or services. Although many things must be evaluated during an EPE, Components like air, water and land are of great importance. Reuse, reduction and recycling are the major components of waste minimization. EPE is usually conducted by the organization on a continuous basis. Since it is an ongoing process, it is usually most cost-effective for the organization to establish continuous monitoring systems of the most important and variable environmental parameters. When this is not possible, the assessment should be on a very regular basis.

Basically, EPE is a management tool or process for enabling an organization to measure and improve environmental performance on a continuous basis. In other word, it relates to minimizing the organization's impact on the environment. A very critical part of the EPE process is the selection of appropriate indicators. If the indicators are selected carefully, the EPE will have more meaning and results in positive environmental improvement. Indicators should be established to cover all systems including management, operational and state of the environment. The collected data can be analyzed in relation to the environmental indicators. So that numerical values could be assessed for each indicator in reference to specific site or product.

Environmental Auditing (ISO – 14010) is usually done on infrequent pattern. On the other hand, EPE (ISO – 14031) is on an ongoing basis involving continuous environmental monitoring. Thereafter, EPE provides a tool to assess the performance of an organization and help to minimize environmental impacts. In other world EPE improves overall management and operational performance and is therefore mandatory for good business.

Since EPE is an analysis that affects everyone and everything, all living organisms have a stake in maintenance of environmental quality. However, EPE is not a strict regulatory requirement or a mandatory standard. It is an ongoing responsibility for all of us.

Different Acronyms Used for Indicators in Different Areas of the EPE

Areas	*Indicator*	
Management	Environmental Management Indicator	EMIs
Operational	Environmental Performance Indicators	EPIs
Environmental State	Environmental Indicators	EIs

Examples of Selective Indicators for EPE

EMIs	*EPIs*	*EIs*
Actions performed by the management employees under procedures, plans and tracking. – Training plants	Assessment of – Actual Discharges – Wastes generated	These indicators help to assess the status of Plants/ animal life including water, air and other resources.
– Legal requirements – Resources and allocation – Control system & their design	– Discharge oremission – Resources used – Actual facility – Equipments for production or services	– Resource impacts – Ecological impacts – Human Health The indicators must cater to the local, regional and global environments.
– Setting of goals – Documentation – Tracking of parameters & reults	– Energy inputs – Heat generated	
– Performance prediction – Root cause identification – Remediation – Corrective measures yetc.		

ISO 14000 Series

ISO 14001	Environmental Management Systems
ISO 14010	Environmental Auditing
ISO 14020	Environmental Labeling
ISO 14031	Environmental Performance Evaluations
ISO 14040	Life Cycle Assessment.

The state of the environment is closely linked to the operational and management components. In reality, the activities in the management and operational systems determine the level of impact that will be generated on the environment. ISO 14031 is the international guidelines that covers EPE and forms the component of ISO 14000 series.

CHAPTER 24
Monitoring of Man Made Pollutants

Vasdeo Singh
Chemical Engineering Department
Faculty of Technology & Engineering
M.S. University of Baroda, Vadodara - 390 001

Abstract

With rapid industrialization pollution problems are also increasing rapidly. Air is an unconfined medium and, therefore, its pollution if not controlled, can be dangerous. The air-pollution problems starting from the source of pollutant to the receptor via pollution control process, dispersion by stack and atmospheric transmission has to be monitored for the protection against any damage to human-life, ecosystems, and cultural assets. The pollutant sources may be industry, transportation and natural causes. The concentration and nature of pollutants may be changed in the atmosphere due to the presence of other materials, humidity and radiations and also due to meteorological factors. Therefore, the monitoring of air pollutants involves not only the sampling and analysis of the pollutants but also the meteorological data monitoring, understanding of atmospheric dispersions and chemical reactions. The quality assurance for monitoring the emissions is very important because the data acquitted is employed for enforcing compliance of standards for air quality control for long term impacts.

Introduction

Air pollution is a multidisciplinary subject and has attracted attention of engineers, physicians, physists, chemists, toxicologists, meteorologists, agronomists, lawyers, etc. The increasing human population resulting in increased industrialization is creating more pollution problems. Air is an unconfined fluid and therefore, its pollution can assume catastrophic dimensions. Thus monitoring the pollutants in the air to control its quality is extremely important for the well being of humans, plants and animals and also the prevention of damage to cultural assets and structures.

Air Pollutants

Air pollutants are generally of two types

Natural

Volcanic gases and ash, dust due to wind, smoke and gases from forest fires, gases and odours from natural decomposition, natural radioactivity , ozone from lightening and the ozone layer, esters and terpenes from vegetation, pollens and other aeroallergens. Over this type of pollution there can be hardly any control.

Man-made

Created by man for power generation, transportation and various industries.

Air Pollutants are also classified as to the origin and state of matter.

1.Origin

(a) Primary – pollutants discharged to the atmosphere from a process.

(b) Secondary – formed in the atmosphere due to chemical reactions.

2.State of matter

(a) Gaseous – SO_2, NO_X, O_3, CO, halogens, HCN , NH_3, metallic organic compounds, metallic carbonyl compounds, various other organic compounds etc.

(b) Particulates (aerosols) – smoke, fog, dust etc.

Sources of Pollutants

There are two types of sources.

(a) Stationary : Various industries

(b) Mobile: Transportation

Objectives of Monitoring

Air pollutants are discharged into the atmosphere by various sources at different heights. The amount and the time of emissions are very uncertain. The concentration of these emissions at a particular location depends on emission sources and meteorological factors. Thus an air pollution monitoring system is required to manage the air quality. The objectives of this system may be –

1. To know the pollutant, its concentration and its sources (if presently unknown).
2. To check the effectiveness of the control techniques provided for its control.
3. To determine the future trends in levels of emissions that will result from new industries or expansions of present industries or residential and commercial developments.
4. To detect the pollutant levels increasing the prescribed limits and the system acting as a warning system to avoid any damage to life or property.
5. To study the effect of meteroligal factors on the transport of air pollutants.
6. To estimate the effects of pollutants on the health, structures, atmospheric processes and visibility.

Monitoring Systems

The parameters monitored are type and concentration of pollutants, meteorological variables, effects of pollutants on health of animals, plants, structures, corrosion, etc.

The scale of monitoring may be local (micro), urban meso (air basin), regional, continental or global.

The monitoring system may be stationary, mobile (in a van, aeroplane or helicopter) or remote sensing.

The number of pollutants to be monitored depends on their chemical properties and dispersion behaviour in the atmosphere. No guidelines can be laid down for this.

Monitoring Network

It comprises of monitoring – the pollutant at the sources and at the pollution control equipment (the concentration and also amount of total emission),

- the pollutant at the plume
- the meteorological variables (wind direction, speed, atmospheric stability etc.)
- the damage (dose, GLC monitoring).

Duration of Sampling

Depends on MAC and purpose of monitoring. No general guidelines can be laid down.

Data Display

The air pollution data are recorded as function of location, magnitude and time. The data are shown in the records as location, maximum values, annual means, and number of occurrences of hourly values above a given concentration as a function of the month of the year. Pollution concentration maps may also be constructed.

Quality Assurance in Air Pollution Monitoring

Air pollution monitoring for purposes such as standards compliance must be prescise and accurate. Quality assurance means quality assessment plus quality control i.e. the system used to verify the analytical process and the mechanism used to control the errors operate with in prescribed limits.

Objectivity, predetermined criteria for data acceptability and formalized written procedures, documentation and action plans are the essential elements of a quality assurance programs. The basic elements of the monitoring system which require quality assurance are: measurement site location and the character of the region, integrity of sampling line, performance of instruments (response time, linearity of response, noise, interference, sensitivity and stability) and calibration. Quality assurance in air pollution monitoring is extremely important.

Conclusions

Air pollution monitoring involves :

- Sampling and analysis of various types of pollutants at different conditions of concentrations, temperatures and pressures.
- meteorological and receptor damage monitoring.
- control system for warning and to bring the air pollutant levels within allowable limits.

CHAPTER 25

Environmental Audits : As a Planning Tool in an Indian Context

Jaydev Nansey
3, Patel Colony,
Jamnagar

Introduction

In the seventies, it was believed that environment has an infinite capacity to absorb and disseminate all the pollutants discharged by industrialization. This concept was questioned in the west and the industrialized nations (Opschoor & Straaten, 1992). However, in the name of development the "undeveloped" countries did not consider pollution to be an important issue. The First Environment Summit at Stockholm in 1972, was an eye opener and the concept of infinite capacity of the environment was doubted (Caldwell, 1991).

In India, various laws and legislations were enacted for environmental protection following the Stockholm Conference (Trivedi, 1993). All these legislations are framed for environmental protection, through strict pollution discharge limits. Like all protective legislations these laws have penalties for violation and no incentives for pollution control. As a result industries were not encouraged to go for environmental protection.

Considering the above problem it is evident, that a new approach towards preserving the environment is required. Environmental Management rather than environmental protection is the answer. The concept of Environmental Planning was introduced to curtail this gap between the industries and the law enforcing agencies. Planning for environmental protection before commissioning of an industry will help the decision making authorities to incorporate the costs and the benefits of investing in pollution abating equipments. Environmental Impact Assessments (EIA) followed by Environmental Management Plans (EMP) and Environmental Audits (EA) are the tools used by an environmental planner to reduce the pollution burden on the industry, through prior planning.

Environmental Audits (EA)

Environmental Audit is a new approach towards involving Industries in the process of overall Environmental Management. It has been observed, that industries considered the Pollution Control Authorities as adversaries. EA brings both these parties together on the same table. Metier of Environmental Audit is the role it plays as a decision making tool useful to the management. International Chamber of Commerce has defined Environmental Audits as (GPCB, 1992) :

"Management tool comprising a systematic, documented, periodic and objective evaluation of how well environmental organization, management and equipment are performing with the aim of helping to safeguard the environment by

(a) Facilitating management control of environmental practices

(b) Assessing compliance with company policies which would include meeting regulatory requirements.

There are two types of Environmental Audit for any industry, internal and external. Internal Audits are carried out for the management to ensure, that it has accurate, up-to-date information on hand regarding the environmental performances of its units. These audits are now a part of a quality approach to environmental management and can be seen as a part of the overall drive for quality assurance. These not only reassures external interests, but also helps in process of setting and enforcing high internal performance standards. External Audits on the other hand are primarily undertaken to fulfil the requirements laid down by the Pollution Control authorities. These audit records are kept open for public reference.

Americans have been the pioneers in introducing EA, which is mandatory under the regulations. This is more of an compliance auditing required by the regulations. European Commission undertakes these audits to check the site improvement plans of various projects (Elkington, 1990).

Environmental Audits in Indian Context

Environmental Audits were introduced in India in 1992 under the Environment (Protection) Rules of 1986, through the powers conferred upon the Govt of India by Section 6 and 25 of the Environment (Protection) Act of 1986. EA has been made mandatory for the industries, which require consent under the Water (Prevention and Control of Pollution) Act of 1974, the Air (Prevention and Control of Pollution) Act of 1981, or the Hazardous Waste (Management and Handling) Rules of 1989 (MoE & F, 1992). Standard formats have been prescribed for scheduled industries (GPCB,1992) and the audit forms are to be submitted to the State Pollution Control Board. Industries are required to submit audited statements by the end of each financial year.

Industries have shown reluctance in compiling with this procedure in the past. Since, this form requires them to detail the raw materials used, nature of product and waste generated. This could lead to closely guarded technical, as well as process details becoming public. Due to the defective and unclear patent laws, audit statements may be misused. To a great extent this fear is valid. However, it is not fair to out rightly reject the Environmental Audits merely on this reason.

Audited statements as prescribed by the Pollution Control Board needs to be investigation detail. For at times industries take such audits lightly and consider it a mere process of paper work of no importance and hence furnish false data.

EA being a altogether new approach towards environment has not been taken seriously by both the Pollution Control Authorities or the industries. There are very few technical experts, who understand the spirit of EA 'in toto'. Unlike the industrial management cadre, all in the field of environment are either technocrats or scientist. This creates a vast communication gap between the technocrats and the managers, and the EA process looses it importance and essence.

Conclusions

Environmental Audits being a new concept requires to be analysed and adopted as per the requirements of Indian Industries. Industries are reluctant to adopt environmental audits because of the two facts, namely the insecurity from the patent laws and the paper work. In order to overcome the former, two sets of audits may be undertaken one external to be made public and another detailed audit for internal use.

In India, industries have shown willingness to adopt the Energy Audits, but are reluctant to adopt Environmental Audits, just because of a simple reason being Energy Audits gives a direct saving on energy cost, whereas, Environmental Audits have no direct benefits. Further, Energy Audit benefits are short termed as well as long term. Whereas, Environmental Audits have a long term benefit, which cannot be quantified within a short period, giving an added excuse to industries for not going for Environmental Audits.

Industries require to be educated about the advantages of undertaking audits. Internal audits can serve as a tool in monitoring the mismanagement of resources and can be coupled with the material management. Thereby, direct benefits of EA will be highlighted giving the industries an incentive for undertaking the same. The practice of environmental audits should not only be an external requirement, but should also be for internal purpose.

Environmental Audits are currently seen by the State Pollution Control Boards as controlling instruments. This approach towards audits may be changed. Further, the audit may be conducted by a combined team of experts from the Pollution Control Boards, Management and Environmental Consultants. The gap between management and technocrats needs to be filled.

Pollution Control Boards are required to update their data base as a cross reference. Moreover, Industries as well as the Pollution Control Officials should be trained and apprised, as and when new methods and procedures for environmental audits are introduced.

Industries should be given incentives for the audits they carry out and should be asked to furnish a future Environmental Management Plan based on the findings of the EA thereby reducing pollution.

In fine, Environmental Audits instead of being a requirement of the Government, should be the requirement of the Industries themselves. The industries should realise that 'A good Environmental Management is a good management'.

References

Caldwell, L. K., (1991). *International Environmental Policy: Emergence and Dimension.* Affiliated East-West Press Pvt. Ltd, New Delhi.

Elkington, J. (1990). *The Environmental Audit.* WWF-UK, London.

MoE&F, Notification vide G.S.R.329 (E), New Delhi, 13 Mar 92.

GPCB, Environmental Auditing (Unpubl) 1992.

Opschoor H. & Straaten J. (1992). *Sustainable Development: An Institutional Approach.* Ecological Economics, 7, Amsterdam.

Trivedi, R. (1993). *Lecture at CEPT on 'Environmental Protection'* (Unpubl), Ahmedabad.

CHAPTER 26
Environment and How It Affects Human Health

Prakash V. Kotecha
Preventive and Social Medicine Department
Medical College, Vadodara

Introduction

Environment is defined as all of the external factors affecting an organism. These factors may be other living organisms (biotic factors) or nonliving variables (abiotic factors), such as water, soil, climate, light, and oxygen. All interacting biotic and abiotic factors together make up an ecosystem.

Organisms and their environment constantly interact, and both are changed by this interaction. Additionally, environmental factors, singly or in combination, ultimately limit the size that any population may attain. This limit, a population's *carrying capacity*, is usually reached because needed resources are in short supply. Occasionally, carrying capacity may be dictated by the direct actions of other species, as when predators limit the number of their prey in a specific area.

Like all other living beings, humans have clearly changed their environment, but they have done so generally on a grander scale than have other species. Some of these changes-such as the destruction of the world's tropical rain forests to create grazing land for cattle or the drying up of almost three-quarters of the Aral Sea, once the world's fourth-largest freshwater lake, for irrigation purposes—have led to altered climate patterns, which in turn have changed the distribution of species of animals and plants.

Scientists are working to understand the long-term consequences that human actions have on ecosystems, while environmentalists-professionals in various fields, as well as concerned citizens in the United States and other countries—are struggling to lessen the impact of human activity on the natural world.

Understanding the Environment

We have realized that the health depends upon various factors some of them are modifiable and others are not. One single largest modifiable factors happens to be the environment. We need to balance between the benefits as a result of the industrial and scientific revolution on one side and possible health and ecological hazards resulting because of the pollution resulting from these activities.

The science of ecology is the study of the interactions that determine the abundance and distribution of organisms. In other words, ecology attempts to explain why individuals live where they do and why their populations are the sizes they are.

No population, human or otherwise, can grow indefinitely; eventually, some biotic or abiotic variable will begin to limit population growth This basic ecological principle was first established in 1840 by German chemist Justus von Liebig and has been called the Law of the Minimum. From a human standpoint, this means that all of the world's physical resources are in finite supply.

In the 1970s the British scientist James Lovelock formulated the Gaia hypothesis, which has attracted many followers. According to this theory, named after the Greek goddess of the earth, the planet behaves like a single living organism. From a scientific viewpoint, the earth is not a single living organism, but it can be viewed as a single integrated system. The National Aeronautics and Space Administration (NASA), using its expertise in planetary and space sciences, is collaborating with other U.S. Governmental agencies in the use of artificial satellites to study global change. NASA's undertaking, begun in 1991, is called Mission to Planet Earth. This project is part of an international effort linking numerous satellites into a single Earth Observing System (EOS). EOS is designed to increase knowledge of the interactions taking place among the atmosphere, land, and oceans; to assess the impact of natural and human events on the planet; and to provide the data that permit sound environmental policy decisions to be made.

Current Issues

In November 1992 a document entitled *Warning to Humanity* was released. 1500 scientists from around the world, including 99 Nobel laureates, a dozen national academies of science, the Pontifical Academy of Science, and the director general of the United Nations Educational, Scientific and Cultural Organization (UNESCO), signed this alarm. The document was bold and clear, stating, "human beings and the natural world are on a collision course" which "may so alter the living world that it will be unable to sustain life in the manner that we know".

The problems facing the environment are vast and diverse. Destruction of the world's rain forests, global warming, and the depletion of the ozone layer are just some of the problems that will reach critical proportions in the coming decades. Their rates will be directly affected by the size of the human population.

Population Growth

Human population growth may be seen to be at the root of virtually all of the world's environmental problems. Increasingly large numbers of people are being added to the world every year. As the number of people increases, more pollution is generated, more habitats are destroyed, and more natural resources are used up. Even if new technological advances were able to cut in half the environmental impact that each person had, as soon as the world's population size doubled, the earth would be no better off than before.

The Population Division of the United Nations predicts that the 5.63 billion humans alive in 1994 will increase to 6.23 billion in the year 2000, 8.47 billion in 2025, and 10.02 billion in 2050. The UN's estimate assumes the population will peak and stabilize at 11.6 billion in 2200. Others predict the numbers will continue to rise into the foreseeable future, to as many as 19 billion people in 2200.

Although it is true that rates of population increase are now much slower in the developed world than in the developing world, it would be a mistake to assume that the population growth problem is primarily a problem of developing countries. In fact, because larger amounts of resources per person are used in the developed nations, each citizen from the developed world has a much greater environmental impact than does a citizen from a developing country. Conservation strategies that would not alter lifestyles but would greatly lessen environmental impact are essential in the developed world.

Evidence now exists suggesting that the most important factors necessary to lower population growth rates in the developing world are democracy and social justice. Studies show that population growth rates have fallen in areas where several social conditions have been met. In these areas, literacy rates have increased, and women are given economic status equal to that of men and thus are able to hold jobs and own property; also, birth control information is more widely available, and women are free to make their own reproductive decisions.

Global Warming

Like the glass panes in a greenhouse, certain gases in the earth's atmosphere permit the Sun's radiation to heat the earth but retard the escape into space of the infrared energy radiated back out by the earth. This process is referred to as the greenhouse effect. These gases, primarily carbon dioxide, methane, nitrous oxide, and water vapour, insulate the earth's surface, helping to maintain warm temperatures. Without these gases, the earth would be a frozen planet with an average temperature of about –18°C (about 0° F) instead of a comfortable 15°C (59° F). If the concentration of these gases were higher, more heat would be trapped within the atmosphere, and worldwide temperatures would rise.

Within the last century, the amount of carbon dioxide in the atmosphere has increased dramatically, largely because of the practice of burning fossil fuels–coal and petroleum and its derivatives. Global temperature has also increased 1°C (about 1.8° F) within the past century. Atmospheric scientists have now concluded that at least half of that increase can be attributed to human activity, and they have predicted that unless dramatic action is taken, temperature will continue to rise by between 1° and 3.5° C (between 1.8° and 6.3° F) over the next century. Although this may not seem like a great difference, global temperature was only 2.2° C (4° F) cooler during the last ice age than it is presently. The consequences of such a modest increase in temperature may well be devastating. Sea levels will rise, completely inundating a number of low-lying island nations and flooding many coastal cities such as New York and Miami. Many plant and animal species will probably be driven into extinction, agricultural regions will be disrupted, and the frequency of severe hurricanes and draughts is likely to increase.

Depletion of the Ozone Layer

The ozone layer, a thin band in the stratosphere (a layer in the upper atmosphere), serves to shield the earth from the sun's harmful ultraviolet rays. In the 1970s, scientists discovered that the layer was being attacked by chlorofluorocarbons (CFCs), chemicals used in refrigeration, air-conditioning systems, cleaning solvents, and aerosol sprays. CFCs release chlorine into the atmosphere; chlorine, in turn, breaks ozone down into its constituent parts of oxygen. Because chlorine is not affected by its interaction with ozone, each chlorine molecule has the ability to destroy a large amount of ozone for an extended period of time.

The consequences of the depletion of the ozone layer are dramatic. Increased ultraviolet radiation will lead to a growing number of skin cancers and cataracts and also reduce the ability of people's immune systems to respond to infection. Additionally, the growth rates of the world's oceanic plankton, the base of most marine food chains, will be negatively affected, perhaps leading to increased atmospheric carbon dioxide and thus to global warming. Even if the manufacture of CFCs was immediately banned, the chlorine already released into the atmosphere would continue to destroy the ozone layer for many decades. Additionally, the latest studies suggest the global warming may increase the amount of ozone destroyed.

Predicting the rate of ozone depletion is difficult. Optimists claim that if international agreements for the phasing out of ozone-depleting chemicals agreed to in Montreal in 1987 are followed, ozone loss will peak in the year 2000. With many of the world's fastest growing countries in the process of industrializing and modernizing, there is reason to believe that destruction will continue to increase well beyond that year.

Air Pollution

Definitions as given in Indian Act

(a) "Air pollution" means any solid, liquid or gaseous substance (including noise) present in the atmosphere in such concentration as may be or tend to be injurious to human beings or other living creatures or plants or property or environment.

(b) "Air pollution" means the presence in the atmosphere of any air pollutant.

A significant portion of industry and transportation is based on the burning of fossil fuels, such as gasoline. As these fuels are burned, chemicals and particulate matter are released into the atmosphere. Although a vast number of substances contribute to air pollution, the most common contain carbon, sulfur, and nitrogen. These chemicals interact with one another and with ultraviolet radiation in sunlight in various dangerous ways. Smog, usually found in urban areas with large numbers of automobiles, is formed when nitrogen oxides react with hydrocarbons in the air to produce aldehydes and ketones. Smog can cause serious health problems. When sulfur dioxide and nitrous oxide are transformed into sulfuric acid and nitric acid in the atmosphere and come back to earth in precipitation they form acid rain. Acid rain is a serious global problem because few species are capable of surviving in the face of such acidic conditions. Acid rain has made numerous lakes so acidic that they no longer support fish populations. Acid rain is also thought to be responsible for the decline of many forest ecosystems worldwide. Germany's Black Forest has suffered dramatic losses, and recent surveys suggest the similar declines are occurring in throughout the eastern United States.

Some of the common air pollution indicators and the sources are given below in the tabular forms for your ready reference.

Common Air Pollutants and their sources

Pollutant	*Major Sources*	*Comments*
Carbon monoxide (CO)	Motor-vehicle exhaust, some industrial processes	Health standard : 10 μg/m³ (9 ppm) over 8 hr; 40 μg/m³ over 1 hr (35 ppm)
Sulfur dioxide (SO_2)	Heat and power generation facilities that use oil or coal containing sulfur, sulfuric acid plants	Health standard : 80 μg/m³ (0.03 ppm) over 1 year; 365 μg/m³ over 24 hr (0.14 ppm)
Particulate matter	Motor-vehicle exhaust; industrial processes; refuse incineration; heat and power generation; reaction of pollution gases in the atmosphere	Health standard : 50 μg/m³ over a year; 150 μg/m³ over 24 hr; composed of carbon, nitrates, sulfates, and many metals including lead, copper, iron and zinc.
Lead (Pb)	Motor-vehicle exhaust; lead smelters; battery plants	Health standard : 1.5 μg/m³ over 3 months
Nitrogen dioxide (NO_2)	Motor-vehicle exhaust; heat and power generation; nitric acid; explosives; fertilizer plants	Health standard : 100 μg/m³ (0.05 ppm) over a year; reacts with hydrocarbons and sunlight to form photochemical oxidants
Ozone (O_3)	Formed in the atmosphere by reaction of nitrogen oxides, hydrocarbons, and sunlight.	Health standard : 235 μg/m³ (0.12 ppm) over 1 hr

Water Pollution

Estimates suggest that nearly 1.5 billion people lack safe drinking water and that at least 5 million deaths per year can be attributed to waterborne diseases. Water pollution may come from point or nonpoint sources. Point sources discharge pollutants at specific locations—from, for example, factories, sewage treatment plants, or oil tankers. The technology exists for point sources of pollution to be monitored and regulated, although political factors may complicate matters. Nonpoint sources—runoff water containing pesticides and fertilizers from acres of agricultural land, for example—are much more difficult to control. Pollution arising from nonpoint sources accounts for a majority of the contaminants in streams and lakes.

With almost 80 per cent of the planet covered by oceans, people have long acted as if those bodies of water could serve as a limitless dumping ground for wastes. Raw sewage, garbage, and oil spills have begun to overwhelm the diluting capabilities of the oceans, and most coastal waters are now polluted. Beaches around the world are closed regularly, often because of high amounts of bacteria from sewage disposal, and marine wildlife is beginning to suffer.

Groundwater Depletion

Water that seeps through porous rocks and is stored beneath the ground is called groundwater. Worldwide, groundwater is 40 times more abundant the fresh water in streams and lakes, and although groundwater is a renewable resource, reserves are replenished relatively slowly. In the United States, approximately half the drinking water comes from groundwater. Presently, groundwater in the United States is being withdrawn approximately four times faster than it is being naturally replaced. The Ogallala Aquifer, a huge underground reservoir stretching under eight states of the Great Plains, is being drawn down at rates exceeding 100 times the replacement rate, suggesting that agricultural practices depending on this source of water may have to change within a generation. When groundwater is depleted in coastal regions, oceanic salt water commonly intrudes into freshwater supplies. Saltwater intrusion is threatening the drinking water of many areas along the Gulf and Atlantic coasts.

The EPA has estimated that, on average, 25 per cent of usable groundwater is contaminated, although in some areas as much as 75 per cent is contaminated. Contamination arises from leaking underground storage tanks, poorly designed industrial waste ponds, and seepage from the deep-well injection of hazardous wastes into underground geologic formations. Because groundwater is recharged and flows so slowly, once polluted it will remain contaminated for extended periods

Habitat Destruction and Species Extinction

It is difficult to estimate the rate at which humans are driving species extinct because scientists believe that only a small percentage of the earth's species have been described. What is clear is the species are dying out at an unprecedented rate; minimum estimates are at least 4000 species per year, although some scientists believe the number may be as high as 50,000 per year. The leading cause of extinction is habitat destruction, particularly of that world's richest ecosystems—tropical rain forests and coral reefs. At the current rate at which the world's rain forests are being cut down, they may completely disappear by the year 2030. If growing population size puts even more pressure on these habitats, they might well be destroyed sooner.

Since European colonization, North America has been transformed: Approximately 98 per cent of tall-grass prairies, 50 per cent of wetlands, and 98 per cent of old-growth forests have been destroyed. This loss is critical from several perspectives. The economic value of species lost and of natural products and drugs that never will be discovered or produced is incalculable. Similarly, it is impossible to place either a moral or an aesthetic value on our growing list of extinct species. As habitats are destroyed and species lost, the world is increasingly losing threads from the interconnected fabric of life.

Chemical Risks

Pesticide residues on crops and mercury in fish are examples of toxic substances that may be encountered in daily life. Many industrially produced chemicals may cause cancer, birth defects, genetic mutations, or death. Although a growing list of chemicals has been found to pose serious health risks to humans, the vast majority of substances have never been fully tested. In recent studies, a wide range of chemicals has been found to mimic estrogens, the hormone that normally controls the development of the female reproductive system in a large number of animal species. Preliminary results indicate that these chemicals, in trace amounts, may disrupt development and lead to a host of serious problems in both males and females, including infertility or increased mortality of offspring, and behavioural changes such as increased aggression. Numerous studies have found that the amount of sperm produced by men has decreased precipitously over the past 50 years.

Environmental Racism

Studies have shown that not all individuals are equally exposed to pollution. For example, toxic waste sites are more prevalent in poorer countries, and the single most important factors in predicting the location of hazardous-waste sites in the United States is the ethnic composition of a neighbourhood. Three of the five Iargest commercial hazardous waste landfills in America are in predominantly black or Hispanic neighbourhood, and three out of every five black and Hispanic Americans live in the vicinity of an uncontrolled toxic waste site. The fact tht the wealth of a community is not nearly as good a predictor of hazardous-waste locations as is the ethnic background of the residents reinforces the conclusion that racism is involved in the selection of sites for hazardous-waste disposal.

Environmental racism takes international forms as well. Dangerous chemicals banned in the United States often continue to be produced and shipped to developing countries. Additionally, the developed world has shipped Iarge amounts of toxic waste to developing countries for less-than-safe disposal.

Energy Production

The world cannot continue to rely on the burning of fossil fuels for much of its industrial production and transportation. Fossil fuels are in limited supply; in addition, when burned they contribute to global warming, air pollution, and acid rain.

Many because of the massive devastation an accident can cause oppose nuclear energy, as an alternative. The accident at the Chernobyl nuclear power plant in 1986 scattered radioactive contamination over a large part of Europe. Approximately 135,000 people were evacuated, and human health has been dramatically affected. The World Health Organization released a report in late 1995 attributing the "explosive increase" in childhood thyroid cancer in Belarus, Ukraine, and Russia directly to the accident.

One reasonable solution is to combine conservation strategies with the increased use of solar energy. The pricé of solar energy relative to traditional fuels has been dropping steadily, and if environmental concerns were factored into the cost, solar power would already be significantly cheaper. Although it is desirable to have a wider range of energy options, other alternative sources of power (such as wind, geothermal, hydroelectric) are not likely to provide large-scale solutions in the forseeable future.

Do we have hopes?

Global environmental collapse is not inevitable. But the developed world must work with the developing world to ensure that new industrialized economies do not add to the world's environmental problems. Politicians must think of sustainable development rather than economic expansion. Conservation strategies have to become more widely accepted, and people must learn that energy use can be dramatically diminished without sacrificing comfort. In short, with the technology that currently exists, the years of global environmental mistreatment can begin to be reversed.

We all have roles to play. We need to know our role, however small or big that may be and start playing it with enthusiasm and rope rather than getting disappointed to the growing menace and feeling helplessness in the area. If each one of us play own role than the hope will get converted to the reality of beautiful and healthy environment.

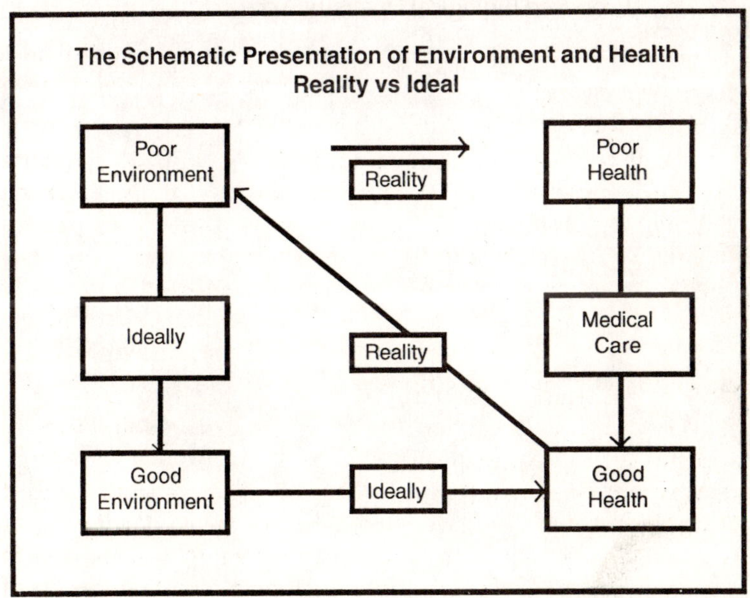

Current Laws in India (Acts and Rules)

1. The Wildlife (Protection) Act, 1972
2. The Forest (Conservation) Act, 1980
3. The Air (Prevention and Control of Pollution) Act, 1981
4. The Environment (Protection) Act, 1996
5. The Chemical Accidents (Emergency Planning, Preparedness and Response) Rules, 1996
6. Rules for the Manufacture, Use, Import, Export and Storage of Hazardous Micro-Organisms and Genetically Engineered Organisms or Cells
7. The Public Liability Insurance Act, 1991
8. The Public Liability Insurance Rules, 1991
9. The National Environment Appellate Authority Act, 1997
10. The National Environment Tribunal Act, 1995
11. The National Environment Appellate Draft Outline of the Proposed Biological Diversity Act.

CHAPTER 27

Role of Monitoring in Air Pollution Control

H.V. Bhavnani*

Deptt. of Civil Engineering, Faculty of Technology & Engg.
M.S. University of Baroda, Vadodara

Industrialisation world over in general and in developing country like ours in particular has resulted in increased material health available to man on one hand and environmental pollution on the other hand. It was the United Nations Conference on Human Environment of Stockholm in 1972 which resulted in the Water and Water Pollution Control Act of 1974 while in 1981 Air Pollution Control Act came in existence. During subsequent years there have been awareness regarding the air pollution and its ill effects but not much have been done to control the pollution. So much so that lately the Courts have started issuing direction to the industries to stop pollution by a fixed date or face closure of the industry.

Pollution control has always been given the last priority as it is nonproductive. Some industries, even though, have installed the pollution control devices, are not using the control devices to save on the running expenses of chemicals, electricity and manpower.

To control pollution, the industries will have to seriously think of changing their attitudes, production technologies, change raw materials to have the effluents and emissions conforming to the pollution standards. This will also need a reliable monitoring of the emissions from the sources and the ambient air.

Presently Exe. Director, A.P. Institute of Environmental Design, Vallabh-Vidyanagar.

The monitoring comprises of two important components (i) Sampling and (ii) Analysis, both of them are equally important. No matter what the analytical instruments or procedures are used, if the sample is not obtained meaningfully most accurate analysis– will have no significance. The importance of monitoring should be appreciated in the control of pollution because it is the analysis results on which will depend the extent and type of pollution control device and setting up of new industries.

The sampling will need the proper planning of network of sampling stations, the interval of sampling, the quantity of samples and the handling of the samples during their transport to laboratory for final analysis.

For source stock sampling, care need be taken to locate the sampling point to avoid turbulence. The sampling must be collected fro.n 16 point across the section under isokinetic conditions. The analysis of the samples is very costly. It involves chemicals as well as equipment.

The pollutants can be basically divided into two categories (i) gaseous and (ii) particulate. Some of the instrumental analytical procedures used for measuring gaseous pollutants are:

1. *Conductometric analysis* working on the principle of Electrical conductivity of solution
2. *Coulombometric* analysis working on the principle of quantity of current flowing
3. *Infrared absorption* working on the principle of infrared light (long wave) absorbed by the pollutants.
4. *Flame ionization* working on the principle of amount of electrons given off by ionization of pollutants.
5. *Flame photometry* working on the principle of spectral intensity and frequency of light given off by spraying sample into a flame.
6. *Colorimetry* working on the principle of absorption of short wave, 5600 A° or other wave length by various pollutants and comparing with calibration curve.

In addition to instrumental analysis, wet chemical analysis can also be used.

For particulate monitoring the instruments, normally, used are dustfall Jar, High Volume sampler, Cascate impinger, Electrostatic precipitator and Thermal preceptor. Now a days equipments using laser technology to determine the particulate size and even size distribution of the particles are available.

Thus once we know the standards of permissible limits of various pollutants, the monitoring will give the following information:

1. The type and concentration of pollutants present in the ambient air.
2. The concentration of pollutants from the source.
3. The status regarding the efficiency of the working of the existing pollution control devices in use.

With the above information, we will be able to control pollution by atmospheric diffusion or by engineering control using control equipment.

In the 'Control of Pollution by Diffusion' the pollutants are not reduced but their concentration is brought down to permissible level by dilution and diffusion. The diffusion pollutants are influenced by (i) meteorological parameters such as wind direction, wind speed, temperature gradient, precipitation (ii) topography of the area (iii) chimney height and even (iv) zoning. This may be supplemented by some floor control devices to bring down the GLC (Ground Level Concentration) with in the permissible limits.

In the engineering control, depending upon the type of pollutants, temperature of the fuel gases, standards required for the final effluent, type of pollutants and efficiency of removal of pollutants, the selection of the pollution control equipment are mandatory. The method used for particulate matter are :

1. gravity settlers
2. cyclones
3. scrubbers
4. electrostatic precipitator
5. fabric filters
6. combustion (flare system)

After the control equipment is installed the regular monitoring must be carried out to know the efficiency of the working of the control equipment used.

CHAPTER 28

Ozone Depletion: A Global Problem

B.V. Kamath[1] and V.S. Patel[2]

[1] *Department of Chemistry, Faculty of Science*

[2] *Faculty of Technology and Engineering,*

M.S. University of Baroda, Vadodara - 390 002

Introduction

An envelope of gases called the atmosphere extending to about 1000 km from the ground surrounds the Earth. The atmosphere acts as a working fluid of the earth's heat engine because of its dynamic activity, which serves to distribute the radiant energy, received from the sun. Most of this energy is converted into kinetic energy or heat before it is re-radiated into space.

The troposphere is the lowest layer in the atmosphere and its temperature generally decreases with altitude. The temperature becomes nearly constant at about 15 km from the ground in the tropics and 8 km at the winter pole. This region is referred to as the tropopause. Above it lies the stratosphere which extends up to stratopasue. Tropopause temperature is in the range 190-215 K which may drop to 180 K, while that of the stratosphere depends on the location and the season, but it increases with height reaching a temperature maximum at about 50 km that forms the stratopause.

The air, which make up the atmosphere has nitrogen (7.8%), oxygen (20%), and argon (1%) as the major constituents. The dry air composition has remained almost constant but the amount of water vapour in the atmosphere varies (4% to 0.1 %) as a result of its contact with surface reservoirs and temperature. The minor constituents that are present only in ppm level include carbon dioxide (3.5 ppm), neon (18 ppm) and helium (5 ppm). The only other gas with abundance greater than 2 ppm is ozone. Its abundance varied with altitude reaching a maximum of 12 ppm in the atmosphere at 30 km.

Ozone Layer

Ozone (MW 48) is a three-atom allotrope of oxygen (MW 32). Ozone has delicate sweet odour at low concentrations and indeed its name is derived from the Greek 'ozien' which means 'to smell'. Schoenbin is credited with the discovery and early investigation of the characteristics of ozone. Ozone is a normal constituent of the sea level environment. Average day-time concentrations are in the order of 1-4 ppb by volume. Ozone concentrations, however, can and do vary as a function of season; man-caused atmospheric pollution, and unusual atmospheric turbulence. The very low, in fact trace, ozone level in our environment nevertheless has a significant effect upon materials: the small ozone constituent of our atmosphere causes the characteristic cracking of stressed rubber, commonly attributed to weathering. Ozone also has a deleterious effect upon textiles, fabrics, organic dyes, metals, plastics, and paints. Ozone also impairs and breathing when it occurs in smog. In turn, ozone oxidizes pollen vision and other unpleasant aerosols in the air we breathe. The biology of plant life is also influenced by ozone. The tropospheric ozone level has increased over the past century as a result of photochemical reactions involving the air pollutants.

Hartley first proposed the presence of ozone in the atmosphere in 1881, which led to the growing interest in stratospheric chemistry. It became important part of environmental chemistry in the early 1970's when Crutzen and Johnston studied the potential stratospheric effects of supersonic aircraft. The study revealed that human activity could affect the chemistry of the cold, remote region 10-50 km above the earth. Of the greatest concern was the destruction of stratospheric ozone, which forms a thin life-protecting shield – the so called ozone layer – against the harmful solar ultraviolet radiations. Ironically human pollution of the atmosphere tends to decrease ozone in the stratosphere where high concentrations of ozone are desirable and to increase ozone in the troposphere where ozone is undesirable because of its toxicity to plants and animals.

The sun emits quite a large amount of energy in the form of ultraviolet, visible and infrared radiations. Although the visible and infrared radiations are primarily responsible for the existence of terrestrial life, the ultraviolet radiation has a damaging effect on the living cells as a result of the changes it elicits in the nucleic acids. If this radiation were to hit living organisms unattenuated they would soon perish. However, the combination of oxygen, nitrogen, ozone and other molecules in the atmosphere lets almost no harmful ultraviolet radiation through to the surface of the earth. While gases like oxygen and nitrogen absorb radiations of wavelength less than 240 nm, ozone is the only atmospheric gas that absorbs in the near ultraviolet viz. 240-310 nm. Ozone appears to be ideal for protection of nucleic acids since the two substances have very similar absorption spectra. For both the absorption is especially strong near 260 nm and the ozone thus takes away just the radiation that would otherwise be absorbed by nucleic acids. Many other important substances like proteins are also affected by these highly energic photons.

The amount of ozone, so important for the continued existence of life on this planet, is very minute. If all the ozone of the atmosphere were forced down to form a layer on the surface of earth, this layer would merely be 5 mm thick. The amount of ozone in the atmosphere is set by the balance between the reactions by which it is formed and those by it is destroyed, in a reaction sequence first proposed by Chapman, now called Chapman mechanism.

Ozone formation by photolysis of oxygen molecules:

$$O_2 \cdot hv \longrightarrow O + O \quad (y < 242\ nm)$$
$$O + O_2 + M \longrightarrow O_3 + M \quad (M = O_2, N_2)$$

$$\text{Net}: 2O_2 \longrightarrow O_3 + O$$

Ozone destruction by photolysis and by reaction with an oxygen atom

$$O_3 + \lambda v \longrightarrow O + O_2$$
$$O + O_3 \longrightarrow O_2 + O_2$$

$$\text{Net} \quad 2O_3 \longrightarrow 3O_2$$

Ozone Depletion

The steady state ozone concentration predicted by the Chapman mechanism is too high to explain the current observation of stratospheric ozone. This is because there are many other ways in which odd oxygen can be destroyed. For example, the ozone is catalytically destroyed in a cycle involving oxides of nitrogen ($NO_x = NO + NO_2$) of natural origin or injected directly into the stratosphere by fuel exhaust from high flying aircraft as first proposed by Johnston:

$$NO + O_3 \longrightarrow NO_2 + O_2$$
$$O + NO_2 \longrightarrow NO + O_2$$

$$\text{Net: } O + O_3 \longrightarrow 2O_2$$

Other trace species such as ClO_x, BrO_x and HO_x, can also lead to catalytic ozone destruction. For example, chlorine from chloroflurocarbons (CFCs) could cause catalytic ozone loss

$$Cl + O_3 \longrightarrow ClO + O_2$$
$$O + ClO \longrightarrow Cl + O_2$$

$$\text{Net: } O + O_3 \longrightarrow 2O_2$$

The CFC's do not react in the troposphere and slowly rise above the ozone layer where they photolyse to release chlorine atoms for the above reactions. Adding these reactions pathways that cause loss to Chapman mechanism have improved the agreement between calculated and observed ozone concentrations.

Scientific investigation going back to several decades has confirmed that the life-saving blanket of ozone is thinning out as a result of these ozone-destroying reactions involving human pollutants. Satellite instrument called TOMS (Total Ozone Mapping Spectrometer) has been mapping the annual ebb and flow of ozone layer. What has surprised the Scientists world-over is the abrupt and massive spring time ozone loss over Antarctica that cannot be explained by the known chemical cycles. There is an area where the layer is so sparse that there is virtually a hole, the so called 'ozone hole'. New chemical mechanisms have been proposed that involve heterogeneous reactions of chlorine at the surface of startospheric particles composed of water vapour and nitric acid, called polar stratospheric clouds (PSCs). These reactions are now known to be important not only for the Polar Regions but also for the entire lower stratosphere. The size of the ozone hole has been increasing every year. In 1979, the ozone hole area was 0.7 milion km^2. It has increased to 27.2 million km^2 in 1998. This is as big as North America and as deep as Mount Everest. Scientists have collected samples from the region where the hole occurs and have found high levels of ozone-eating chemicals. No hole has yet been found over the Arctic, although the chemicals that could cause it are present. But the ozone layer has certainly been found to thin out over the Northern hemisphere in general. In a band that stretches around the globe in the latitudes of Canada where Hudson Bay lies, up to 7 per cent ozone depletion has been found in winter.

Ozone-Depleting Chemicals

The best-known ozone-depleting chemicals are chlorofluorocarbons (CFCs). These are used in a variety of aerosols like hair sprays, antiperspirants, insect sprays, and spray paints. CFCs are also used extensively in refrigerators, air conditioning systems and packaging foams. The other chemicals that threaten ozone layer include carbon tetrachloride used in the manufacture of CFCs. Halons used in fire extinguishers, methylchloroform used as a solvent in glues and spirit base pens and trichloroethane used in typist's correction fluid.

Key dates in the History of Ozone-depleting chemicals

Dec 1973	Rowland and Molina made their discovery.
Oct 1978	The use of CFC's in aerosols was banned in USA.
Oct 1984	British team reports 40% loss of ozone over Antarctica during austral (Southern hemisphere) spring.
Sept 1987	Montreal Protocol – representatives from 43 countries agree to CFC reductions by 2000.
Oct 1987	Antarctic expedition verifies huge losses of ozone over the area during austral spring.
Mar 1988	United Stated ratifies Montreal Protocol; large US manufacturers announce they will cease production of CFC's.
Apr 1988	Plastic foam manufacturers announce they will stop using CFC's.
Mar 1989	Seven hundred representatives from 124 countries attend London Conference on saving the ozone layer.
June 1990	Environment Ministers from 93 countries agree to strengthen Montreal protocol with complete phase out of CFC's by 2000 (and HCFC's by 2040).
Oct 1990	US Congress passes revised Clean Air Act that includes phase out of CFC's by 2000.

contd....

Jan 1991	Environment Ministers of European Community agree to complete CFC ban by 1997.
Jan 1992	Increased concentrations of ozone-depleting chemicals are found over populated areas in the Northern hemisphere.
Feb 1992	US President moves target date for phase out of CFC's from the year 2000 to 1995.
Jan 1994	Halon production stops. EPA formally asks some companies to continue production of CFC's through 1995 to meet "consumer needs" in automotive air conditioners.
Oct 1994	Antarctic ozone hole appears earlier than normal covers 23 to 24 million square kilometres.
Jan 1995	NASA satellite data confirm CFC link to ozone hole.

Global Problem

The discovery of 'ozone hole' over Antarctica in 1982 caused great International concern. If ozone depletion is not reversed there could be 60 million more skin cancers by 2075, not to mention the withering of important food crops and the deaths of animals crucial to the food chains. When Scientists finally prove that CFCs posed a clear and immediate danger to the ozone layer, thirty-one nations rapidly banded together and signed a historic agreement in Montreal in 1987 called the Montreal protocol, to begin phasing out the CFCs by the year 2000. Realizing that the problem may lead to planetary collapse, more countries including the developing countries, agreed to phase out the production of all the ozone depleting chemicals. Developing countries have been given longer than industrialised countries to phase out CFCs and related chemicals. One of the strongest inducements to sign was trade restriction that was to be imposed on nations who did not sign. There have been amendments to the protocol, but the real enforcement of restrictions on the manufacture, use and distribution of substances that can deplete the ozone layer is contained in the 1989 agreement. Now it has become vital that all countries, to overcome the governmental inertia and national rivalries and work together to combat this global problem. Even so, chlorine concentrations in the atmosphere are unlikely to return to their pre-CFC level until the end of the next century.

References

Bjöm, L.O. (1976). *Light and Life*, Hodder and Stoughton. London

Development and the Environment. World Development Report 1992. Oxford University Press.

Ingersoll, A.P. (1983). *The Atmosphere*. In Scientific American. September 1983.

Kaku, M. (1998). *Visions*. Oxford University Press.

Macalady, D.L. (1998). *Perspectives in Environmental Chemistry*. Oxford University Press.

Chapter 29
Population Explosion and Pollution

Sheo Kumar
BSI, Kolkata - 700 001

Man has tremendously disturbed and vigorously manipulated the nature for their own benefits since the civilization of mankind. Among which urbanization and industrialization are more important. In India, urbanization is intricately linked with the process of economic development and is thus considered inevitable. The population explosion due to uncontrolled birth rate and decreased death rate has created a great problem for the common man, municipal authorities and government to improve the quality of daily life. At present, 27 per cent of the population is inhabiting in urban area and by 2015 AD the population will be more than 50 per cent and in over 50 cities the number will be greater than one million. Nearly 40 per cent of urban population is below poverty line, out of which 20-30 per cent population is distributed in slum areas where there is no shelter, adequate drinking water supply, proper sanitation, drainage facilities, waste disposal service and health center. The overall quality of urban environment has deteriorated since the beginning of developmental processes in the larger cities reaching at saturation point by massive traffic, air, water and noise pollutions as well as increased unsafe social environment and unable to cope up with the increasing pressure on their infrastructure.

Since 1960, various nature of industries are growing catastrophically in and around urban areas and in future its number will increase immensely for development of nation's economic condition but release of their uncontrolled and untreated wastes and discharges have caused sever air and water pollution. Leakage of methyl-iso-cyanate (MIC) gas from KILLER TANK 610 at Bhopal on 4th December, 1984 had killed over thousands of human beings, animals and damaged a number of plant species in a day. Combustion of fossil fuel, exhausts of thermal power plants, refineries etc. are continuously increasing the concentration of undesirable ambient air pollutants i.e., SO_2, CO, NO_2, NO_X, O_3, SPM etc. in the environment are solely responsible for causes of respiratory, eye and other diseases. Deterioration at marble at "Taj" due to 'acid rain' and depletion in ozone layer by the emission of chloro-fluoro-carbon (CFC) from the refrigerators and conditioners are important developmental tips. The wastewater of tanneries, textile processing industries, distilleries etc. are regularly contaminating the precious water resources i.e., rivers, lakes, reservoirs and other water-bodies by direct disposal of effluents, seepage and percolation and dumping of solid wastes in open area resulted to land pollution by leaching. The overall impact of the existence of industries on urban population and environment is increase of smoke, heat, noise, gas, fumes, liquid and solid wastes not only affecting the human health but also the plants and animals (aquatic and terrestrial) and valuable cultural wealth.

To abate different environmental problems and to maintain biodiversity, developed and developing countries have implemented a series of awareness programmes at global level instead of local level, while rapid deterioration in urban environment and shrinkage in biodiversity is totally dependent upon the residents, industrialists, concerned authorities and policy makers. Besides, the basic resource like potable water is dwindling by increasing air and water pollution, increase of atmospheric temperature due to green house effect, low rate of rainfall, sinking of ground water level etc. Thus, it is imperative to find out the sources of pollution at grass root level with a rigorous analysis of the assimilative and supportive capacities of the environment and needs a consultative process among the urban residents, governments and non-governmental organizations to make the cities more livable. Then only we can achieve the target of Agenda 21 drafted in Chapter 28 in June 1992 during Earth Summit held at Rio de Janeiro, Brazil. Recently, National Institute of Urban Affairs, New Delhi has completed an urban environmental assessment of four Indian cities *viz.*, Delhi, Bombay, Ahmedabad and Vadodara. It is a good start and hope in next few years we will be able to overcome the problem by implementing effective control measures for the welfare of the mankind.

Subject Index

D

E

F

G

H

I

J

K

L

M

V

W

X

Z

Generic Index

T

U

V

X

Z